WITHDRAWN

Einstein as myth and muse

EINSTEIN

AS MYTH AND MUSE

► **ALAN J. FRIEDMAN**
University of California, Berkeley

► **CAROL C. DONLEY**
Hiram College, Ohio

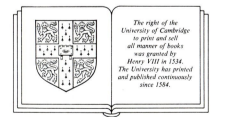

The right of the
University of Cambridge
to print and sell
all manner of books
was granted by
Henry VIII in 1534.
The University has printed
and published continuously
since 1584.

CAMBRIDGE UNIVERSITY PRESS

Cambridge

London New York New Rochelle

Melbourne Sydney

Published by the Press Syndicate of the University of Cambridge
The Pitt Building, Trumpington Street, Cambridge CB2 1RP
32 East 57th Street, New York, NY10022, USA
10 Stamford Road, Oakleigh, Melbourne 3166, Australia

First published 1985

Printed in Great Britain at the University Press, Cambridge

Library of Congress catalogue card number: 84-23839

British Library Cataloguing in Publication Data

Friedman, Alan J.
Einstein as myth and muse.

1. Einstein, Albert – Influence
2. English literature – 20th century – History and criticism
3. Literature and science
I. Title II. Donley, Carol C.
820.9'008   PR478.E3/

ISBN 0 521 26720 X

TP

To Mickey Friedman and Alan Donley

CONTENTS

PREFACE

Observers of the flow of a culture delight in tracing the rapids. Turbulence, intense interaction, and universal excitement may be confined to a matter of years or decades, but those periods often determine the direction for centuries. The philosophy of classical Greece, the Roman expansion, the Renaissance, and the Age of Reason marked periods of rapid change embracing all facets of Western culture. These revolutions have increasingly involved science as that most cumulative of man's enterprises developed its power. In the "atomic age" of the 20th century, science has become a dominant cause of cultural divergence.

Great revolutions in the science of physics occurred in the early decades of the 20th century. The hero of the first revolution was Albert Einstein, whose work and life became central to this century's awareness of science. Einstein's work provided a new description of the physical universe, a world view so radical that its implications are still percolating through society. He also created a new set of fundamental images, additions to the cultural language, that immediately found applications beyond science in fiction, poetry, art, and advertising.

In the first quarter of the century, Einstein's exciting ideas established him as a muse from science, inspiring and supporting experimentation in the arts. Soon the kindly, humane, distracted figure of Einstein became a personal image for the joys of the intellect. That image took a drastic turn with the explosions of the atomic bombs of 1945. Einstein suddenly came to represent a contemporary version of the Prometheus myth, bringing atomic fire to a civilization unprepared to handle its immense powers.

Our study endeavors to examine connections between Einstein, as both myth and muse, and the contemporary cultural enterprise. What made our work so intriguing is that Einstein did not merely move with the flow of cultural history, but cut a new channel across the conventional separations of science and the humanities.

ACKNOWLEDGMENTS

Research for this book was supported in part by a grant from the National Endowment for the Humanities' Basic Research Program. We would like to thank the Endowment and our project officer, Nadina Gardner, for their encouragement and assistance. We wish to thank Robert Karplus of the University of California, Berkeley, who administered the grant for the University, and gave us numerous suggestions on the research and on the manuscript itself. At the University of California, Lawrence Hall of Science chief librarian Priscilla Watson, with Editorial Assistant Valerie Wheat, helped us throughout the research and writing phases. Jill Kangas of the Lawrence Hall provided administrative and secretarial aid throughout the project. The Computer Operations, Business Office and Development Office of the Lawrence Hall, and the Computer Center of the University of California at Berkeley, provided continuing support for this project.

Three institutions helped us locate the general research material on which this book is based: the Doe Library of the University of California at Berkeley, the Niels Bohr Library at the Center for the History of Physics in New York, and the Hiram College Library. We are grateful to the staff of these institutions for their patience and expertise. For their help in obtaining special materials, we wish to thank Dr. Lola Szladits, Curator of the Henry W. and Albert A. Berg Collection of English and American Literature of the New York Public Library, for granting access to Virginia Woolf's letters and diaries; Joan Warnow, Associate Director of the Niels Bohr Library, for granting access to letters and tape recordings of Niels Bohr; Spencer Weart of the Center for the History of Physics, for a draft of his study of images of nuclear power and weapons; and Alva Rogers, for access and guidance through his remarkable collection of early science fiction and fantasy.

We also wish to thank colleagues including Donald Dooley, Emeritus Professor of Physics at Hiram College, who read an early draft of Carol Donley's dissertation and who alerted her to the work of Gerald Holton; Robert Bertholf, Bruce Harkness, Richard Madey, Martin Nurmi, and Bob Smith from Kent State University, Eric Rabkin from the University of Michigan, and Andrew Fraknoi of the Astronomical Society of the Pacific, all of whom gave critical assistance and encouragement; and Hardin Goodman of Florida State University, who was the first inspiration for Alan Friedman's investigations into the relations between science and literature. Fellow members of the Division of Literature and Science of the Modern Language Association gave us continuing support and critical aid.

Readings and criticism of the full manuscript have been invaluable to us, and we would like to thank Eric Rogers, Arthur I. Miller, and Marie R. Galbraith for their detailed comments and suggestions. We would also like to thank Simon Mitton and the staff of the Cambridge University Press for their work on the publication.

"War is Kind", from *The Collected Poems of Stephen Crane*, published by Alfred A. Knopf, Inc.

"St. Francis Einstein of the Daffodils," from William Carlos Williams, *Collected Earlier Poems*. Copyright 1938 by New Directions Publishing Corporation. Reprinted by permission of New Directions.

Permission to reprint selections of articles in the *William Carlos Williams Review* (4/1, Spring 1978, pp. 10-13 and 5/1, Spring 1979, pp. 6-11) granted by editor Peter Schmidt.

Permission to reprint material from Carol Donley's "Einstein's Influence on Modern Poetry," in *After Einstein*, ed. Peter Barker and Cecil Shugart (Memphis State University Press, 1982) granted by J. Ralph Randolph, director.

Fig. 1 (cartoon from *The New Yorker*) drawing by Rea Irvin; © 1929, 1957, by The New Yorker Magazine, Inc., is reproduced by permission.

Fig. 2 (cluster of galaxies in Hercules) is photograph number 2303, © 1973 National Optical Astronomy Laboratories/Kitt Peak, and is reproduced by permission (Agnes Paulsen, Public Information Office, NOAO).

Figs. 3 and 15 (*Time* covers) are © 1946 and © 1979 Time Inc. All rights reserved. Reprinted by permission from *TIME*.

Fig. 4 (detail from the *Pittsburgh Post-Gazette*) is reproduced by permission, courtesy of the American Institute of Physics, Niels Bohr Library.

Fig. 5 (U.S. postage stamp of Einstein) is © U.S. Postal Service, and is reproduced by permission.

Figs. 7 and 8 (Pennaco Hosiery catalog) are reproduced by permission, courtesy of Barbara Guzy, Pennaco Hosiery Division, International Playtex, Inc.

Fig. 9 (Carlsberg advertisement) has been reproduced courtesy of J. Goodstein, California Institute of Technology.

Fig. 10 (Data General advertisement) is © 1975 by Data General Corporation, and reproduced courtesy of Data General.

Fig. 11 (detail from an Allied Chemical advertisement) is © 1977 Allied Chemical Corporation, and has been reproduced by permission from Allied Corporation courtesy of Mark R. Whitley, Manager, Corporate Advertising.

Fig. 12 (Norsk Data advertisement) is reprinted by permission from Norsk Data S.A, courtesy of Don Pyle.

Fig. 14 (Einstein in 1905) is reproduced by permission of the Hebrew University of Jerusalem, courtesy of the American Institute of Physics, Niels Bohr Library.

Finally, we would like to thank our spouses, Mickey Friedman and Alan Donley, and our families, including Mr. and Mrs. George Friedman, for the endurance and support which helped us throughout this project.

INTRODUCTION

Students of literature remember the year 1922 for the publication of James Joyce's *Ulysses* and T. S. Eliot's *The Waste Land*. Musicians remember it for Arnold Schoenberg's development of twelve-tone serial music. Also in 1922, Albert Einstein won the Nobel Prize in physics. That these major events in the intellectual history of our era happened concurrently is no coincidence. The physics, art, and literature of 1922 differed markedly from the traditions those fields had accepted for several preceding generations. Revolutions were occurring which went beyond the usual temporary revolts of a younger generation against the rules of its elders. In less than three decades, these revolutions overthrew metaphysical and epistemological assumptions which had been deeply ingrained in human thinking for centuries.

Twentieth-century physics postulated and experimentally validated two new world views, relativity theory and quantum theory, which differed from the conventional so fundamentally that philosophers and artists were encouraged to assimilate similar revolutionary views into their own disciplines. Modern painting, for example, developed a new spatial ordering. The new world views opened a huge realm of unexplored possibilities and demanded a change in our imaginative picture of the world.

Artists in all fields welcomed this opportunity, studied the new ideas, and experimented with art forms consonant with their time. Wyndham Lewis points out that the experiments involved "not only technical and novel combinations, but also the essentially new and particular mind that must underlie, and should even precede, the new and particular form, to make it viable."[1] That new and particular mind developed from the interaction between disciplines dealing with the new world views. Not since the Renaissance had so many creative people paid attention to what was going on in other fields.

Many artists contemporary with physicists Albert Einstein, Werner Heisenberg, and Niels Bohr found that conventional forms were inadequate in the context of the new physical understandings. New forms had to be created in order to give shape to the new ideas, and the new forms in one discipline often paralleled those in others. Thus as four-dimensional geometries appeared in mathematics and physics, treating time as a dimension equivalent to those of space, the traditional three-dimensional representation in art expanded to include temporal dimension in avant-garde painting. The arts became concerned with time as a dimension and with space-time relationships. Physicists found there was no universal frame of reference, and multiple viewpoints showed up on paintings and in novels. Physicists indicated that the flux of a field represented the strength of its source, and artists began inventing their own "field" forms. Relationships between objects became more important than the objects themselves. When physicists discovered they could not simultaneously find a particle's exact position and exact momentum, they admitted the basic indeterminism inherent in the quantum theory, which gives a probabilistic rather than a causal description of nature. In literature conventional consecutive plots often disappeared and standard syntactical relationships were fractured.

The interactions between diverse disciplines characteristic of this revolution owed much to modern technology and communications. Widely disseminated newspapers and magazines made headlines out of each new advance in science and the arts. Collections of the new art traveled quickly across continents and oceans. Creative people in all disciplines moved from one center of activity to another, working in an atmosphere of enthusiastic collaboration, and making the revolution truly international and interdisciplinary. Schoenberg's twelve-tone serial music was heralded in the same cultural moment as Picasso's cubism, Calder's mobiles, Einstein's relativity, and Joyce's *Ulysses*. The similarities of themes and experimental images, as well as the common cultural milieu, suggest that the new art forms did not develop in isolation; they carried the ideas of a broader, interdisciplinary new world view which they were created to express.

Twentieth-century literature grew up in the same environment with relativity, quantum theory, process philosophy, and modern art. They all created and shared a family of ideas carried by new forms. From the conventional viewpoint, these new forms often contradicted common sense. Those unwilling to accept the new world views often expressed outrage and disgust over the new forms. Riots frequently occurred at exhibits and performances of modern works, mainly because the rioters did not understand what was going on. Comparable vehemence accompanied Einstein's entry into public awareness. Even those making light of the new forms were aware of possible interrelationships between disciplines.

> In a notable family called Stein,
> There were Gertrude, and Ep, and then Ein.
>> Gert's writing was hazy,
>> Ep's statues were crazy,
> And nobody understood Ein.[2]

For many who did understand, the new concepts brought "a springtime of the mind."[3] For them, the new forms in all fields did make sense. As Wyndham Lewis commented, "Point for point what I had observed on the literary, social, and artistic plane was reproduced upon the philosophic and theoretic: and with a startling identity, the main notion or colour at the bottom of the theoretical system was precisely the same thing as what was to be observed throughout the social and literary life of the day."[4]

To illuminate some of the parallels in these cultural revolutions, we will examine two dramatic themes: the reshaping of physics by Albert Einstein and his colleagues, and the changes in literature in the 20th century. The thesis of *Einstein as Myth and Muse* is that science does have a deep influence on other aspects of culture, even those that may seem far removed, such as "serious" literature. Major contemporary writers have been implicitly and explicitly affected by ideas in fundamental science, in particular by Einstein's theory of relativity and by quantum theory, even when those ideas have been filtered and distorted by the ubiquitous popularizations. Attention to these revolutions in science helps explicate and enrich understanding of works by such seminal authors as William Carlos Williams, Virginia Woolf, Vladimir Nabokov, and William Faulkner. The overwhelming preponderance of criticism of these works makes no mention of science, so that our work may improve the understanding of these particular authors, as well as the broader issues of the relations between science and the humanities.

Critics have long considered the impact on literature of the Copernican-Galilean-Newtonian revolution to be a legitimate and fruitful area of critical study. One thinks of the work of Marjorie Hope Nicolson and Douglas Bush in particular (for example, Nicolson's *Newton Demands the Muse: Newton's Opticks and the Eighteenth Century Poets*; and Bush's *Science and English Poetry*). Also most literary critics acknowledge and work with certain influences and relationships between Darwinian evolution and such literary movements as naturalism. Freud's impact on modern literature has been, if anything, overworked. Most critics shy away, however, from considering the impact of Einstein's relativity and the concurrent developments in quantum theory and atomic physics. Usually, either the relationships between modern physics and literature are ignored or they are treated in such a facile and superficial way as to deprive them of useful meaning. To claim, for instance, that any weird event or bizarre set of circumstances takes place in some kind of "fourth dimension" or that a flashback indicates the "flux of a space-time continuum" is simply to play with confusing and thoughtless metaphors. On the other hand, to claim that no positive relationships exist between modern physics and literature or to state that such studies are faddish is to deny a major metaphysical bias of this century.

What is needed and what our study attempts to accomplish is a clear layman's description of the pertinent physics, a historical tracing of how the new ideas became public property and affected the artists, and a careful, accurate use of the physics where it can be legitimately brought to a critical discussion of the literature. Both the physicist and the critic must be involved in this undertaking. One author of this book was trained as a physicist (Alan J. Friedman) and one as a literary critic (Carol C. Donley), but each has studied and written about the other's field. This book is a true collaboration, and there is no chapter which is the exclusive work of either author. We have each written substantial parts of the science popularization and of the literary analysis, and we each had our favorite authors to present. Thus, we have tried to achieve a thoroughly interdisciplinary approach.

We have limited our discussions primarily to physics and literature, not to deny the importance of other cultural movements, but because we are most familiar with these areas, and we have found a wealth of compelling examples of connections between them. In examining the literature, we will concentrate on those aspects that are informed by an awareness of the changes in science, and will not attempt to present complete treatments of authors and their works.

Albert Einstein created the first revolution in 20th-century physics with his theory of relativity and was a crucial force in the second revolution, quantum theory. He was not the only revolutionary in 20th-century science, but beyond the community of scientists and historians of science, Einstein is easily the most discussed of the physical scientists. Thus Einstein came to serve as a personification of the new physics for artists, writers, and the general population. In those cases for which authors or artists have mentioned the influence of science on their works, a newspaper story or essay on Albert Einstein is typically cited as the launching point. Einstein is also the one 20th-century scientist about whom biographical details and anecdotes have entered general public usage. Tales of Einstein's life occur repeatedly in popular literature, and these, no less than Einstein's science, have had a powerful influence on our culture's conceptions of science and scientists. Poetry by William Carlos Williams and Archibald MacLeish will demonstrate the use of Einstein himself as an intellectual hero in the decades following his sudden fame in 1919. The influence of Einstein's dramatically different personal image after World War II will be treated in the final chapter of this book.

Most of the interesting relations between science and literature are not announced through direct quotations or appearances by Einstein, however. More often we will present analyses of literary texts, such as the work of Lawrence Durrell, Vladimir Nabokov, and Louis Zukofsky, which contain citations or language that clearly speak of inspiration and incorporation of ideas from science. We shall also examine texts, including works by Virginia Woolf and Robert Coover, that lack explicit statements documenting connections with science. Some of these works contain deliberate but unannounced and indirect use of science; others incorporate concepts from science gleaned from the climate of thought without specific awareness; a few represent independent invention of patterns and language that parallel inventions of science.

We shall indicate in each case what kind of relation we believe exists between science and the literature discussed. Sometimes concepts in science became distorted as they were transferred from scientists to non-scientists, and in these instances we shall compare the various versions.

Other contemporary critics are striving to examine the nature of such cultural interaction on a more general theoretical level. "Post-modern" criticism offers many tools for such analysis, including structuralist, de-constructionist, and epistemological approaches. The works of Michel Foucault and Jacques Derrida suggest the importance and fascination of such examinations, and the promise that general cultural theories may reveal underlying fibers that connect the specific examples of influence that we offer.[5] We do not try to propose global interpretations, but we do hope we provide new data for cultural theorists to use.

Our explanations of twentieth-century science and its history are intended to aid readers who lack background in these areas. The history of science is at present undergoing some exciting developments. New forms and methods of analysis have been tried, and controversy has arisen about several of these. We have not tried here to undertake any radical departures in our presentation of science and its history. We did find the analyses of Thomas Kuhn to be helpful because they examine "revolutionary" changes in science, and the journalists, novelists, poets and dramatists we discuss almost all regarded twentieth century science as revolutionary. However, many historians of science believe the emphasis on scientific change as revolutionary has been overstressed, and call attention to evolutionary, thematic, or other models of development in science. Thus we cite historians with differing views, and do not claim that the "paradigm" revolutions of Kuhn are the only useful way to look at twentieth-century physics.

Herbert Muller encourages interdisciplinary endeavors such as these for their potential contribution to the definition of fundamental concepts: "The all-important implication of the physicists' new view of the world is that it makes possible a reconciliation of philosophy, art, and science after long years of invidious specialization in different orders of 'knowledge' and exasperated dispute over different levels of 'truth'."[6] That reconciliation began in the first three decades of this century and yielded extremely fruitful results. Ignoring these relationships necessarily limits the critical approaches to much modern literature. Paying attention to these relationships, on the other hand, opens up new areas for critical study and interpretation.

1

The popularization of the new physical ideas

Scientists leave their discoveries, like foundlings, on the doorstep of society, while the stepparents do not know how to bring them up.
Alexander Calder

Anyone studying the revolutionary first three decades of the 20th century must be careful not to view that time only from the perspective of the present. An outrageous idea seventy years ago may be a conventional one today. The public now takes for granted a pace of scientific discovery and new concepts that astonished people at the turn of the century. For example, doctors and dentists now use X-rays routinely for diagnosis. It is easy to forget that Wilhelm Roentgen did not discover X-rays until 1895 and that the public was so fascinated with these new pictures that it crowded into special galleries and exhibits in cities all over the western world to view simple X-rays of hands and teeth.

Soon after Roentgen's discovery, Henri Becquerel discovered radioactivity and began a whole new study of radiation. The physical world suddenly appeared full of bizarre occurrences – of "indivisible" atoms breaking down spontaneously, of high-energy rays penetrating solids, of atoms that were mostly empty space. The medical advantages of X-rays and radioactive elements were overshadowed by the wonder of the curious new atomic world they revealed and of humanity's relationship to it.

Other new sciences were born at the same time. Although Gregor Mendel discovered the basic laws of heredity in the 1860's, his paper on the work lay undiscovered until the early 1900's when scientists first recognized what genes and chromosomes were. Suddenly the science of genetics captured the public imagination. Scientists and laypeople alike developed plants with such desirable qualities as disease resistance, higher crop yields, and shorter growing seasons. Animal husbandry became a science. The old taboos against incest and marriage between close relatives were finally validated by scientific findings. Recessive and dominant genes, mutations, hybrids, and crosses became part of the public vocabulary as the old Aristotelian theory of "blood lines" gave way to the new genetics. Genetics also spurred renewed interest and credibility for Darwin's theory of evolution. Genetics provided a mechanism by which living organisms could change over the ages and pass those changes on to succeeding generations.

Scientists trying to relate the physiology of the nervous system to human behavior developed the new science of psychology. Freud's major works all came into the public arena in the first three decades of this century. Psychoanalysis became a medical tool for the treatment of mental illness. People began examining the subconscious, analyzing dreams, talking about the id, ego, and superego, discussing complexes, inhibitions, and drives as the new psychology became the subject of everything from cocktail party conversations to full-length novels.

With the aid of more powerful telescopes, astronomers discovered whole galaxies of stars outside our own Milky Way and showed that the universe is expanding as if it were born in a cosmic explosion. The sun is far from the center of the Milky Way galaxy, and the Milky Way is not the center of the universe, if such a center even exists. Humankind had come a long way from the central position in biology and space it had occupied in the Aristotelian universe.

At the same time, advances in technology produced the automobile and the airplane, electric light and indoor plumbing, radio and motion pictures, all within about fifty years. Many of these inventions not only made life easier but also greatly increased communication so that word of innovation in any field found a wide audience.

▶   *The public awareness of change*
Because science and technology were making such a remarkable difference in people's lives, they attracted a great deal of public interest. That interest did not focus only on novel gadgets. Instead of solving the last few problems in a mechanical universe, science had opened up a Pandora's box of new world views whose contents required entirely new approaches and attitudes. Public discussion began to wrestle with the new concepts of relativity, of indeterminism, of an expanding universe, of the hypothetical – rather than absolute – nature of scientific laws. These revolutionary ideas pervaded the western world as part of the daily fare of newspapers.

According to one part of the new physics (quantum theory), truly isolated entities cannot exist even in principle. An analogous postulate certainly holds true for the new theories themselves which by no means remained trapped in laboratories or isolated in professional journals. As might be expected, the further the revolutionary ideas spread into the non-scientific world, the more the distortion and confusion of those ideas increased, not only because of their fundamental novelty and the difficulty of translating them into non-mathematical language but also because most translators, the journalists and educators, did not themselves have the thorough mathematical and scientific training needed to understand the original scientific publications.

Widely believed misconceptions arose: that Einstein taught everything is relative, including truth; that all observations are subjective; that anything is possible. The lessons of relativity were subtler, stranger, and more authoritarian than these (see Chapter 3). Yet, however distorted the ideas became, this revolution of the new physics sent its shock waves into many non-scientific fields and contributed to corresponding changes in the arts and humanities.

Physicists explaining their own positions in articles and books usually achieved the most accurate interpretations of the actual physical theories, although their explanations were not always accessible to the general public. A few scientists turned spontaneously into philosophers (often rather poor ones) who published their metaphysical, epistemological, and ethical interpretations. James Jeans and A. S. Eddington, for example, both claimed that the new physics gave free will to humanity. Such claims, which many physicists and philosophers criticized, were nevertheless repeated by a number of laypeople who joined the publishing rush.[1] Some popularizers gave excellent explanations of the new physics, but others felt inspired to construct whole ethical systems, to support dialectical materialism, or to give scientific justifications of spiritualism – all on the basis of the new ideas.

The philosophers Henri Bergson, Bertrand Russell, and Alfred North Whitehead, who paid close attention to the new developments, wrote extensively on the modern physical theories and their relationships with philosophy. Russell and Whitehead had the advantage of being excellent mathematicians and logicians, while Bergson's own philosophy had at least metaphorically anticipated, and perhaps even influenced, some of the physical theories.

Indirectly analogous world views were held by avant-garde artists, many of whom had anticipated elements of the physics revolution and gladly absorbed congenial aspects from the new theories as they appeared.

▶   *Einstein on the front page*

Except for isolated avant-garde groups, the general public remained untouched by Albert Einstein's revolution in physics through World War I. This situation changed dramatically, however, when the results of the British eclipse expedition (discussed in Chapter 3) were announced in November 1919. The London *Times*, in an article headed "The Fabric of the Universe," explained the Eddington expeditions and their purpose. The article concluded: "It is confidently believed by the greatest experts that enough has been done to overthrow the certainty of ages, and to require a new philosophy, a philosophy that will sweep away nearly all that has been hitherto accepted as the axiomatic basis of physical thought."[2]

The Times was not the only major newspaper to make so much of the relativity theory and its apparent verification by the Eddington expedition. *The New York Times* of Sunday, November 9, carried a three-column article headed, "Eclipse Showed Gravity Variation: Hailed as Epochmaking." After describing the expeditions, the paper stated, "The evidence in favor of the gravitational bending of light was overwhelming, and there was a decidedly stronger case for the Einstein shift than for the Newtonian one."[3] The article then quoted J. J. Thomson's assessment of the experimental verification of relativity. "It is not a discovery of an outlying island, but of a whole continent of new scientific ideas of the greatest importance to some of the most fundamental questions connected with physics."[4] *The New York Times* did not cease its coverage after one news item. The next day brought another article, this one headed, "Light All Askew in the Heavens: Einstein's Theory Triumphs."[5] On November 11, a tongue-in-cheek editorial commented, "This news is distinctly shocking, and apprehensions for the safety of confidence even in the multiplication table will arise."[6] The following Sunday, *The New York Times* carried two articles and an editorial devoted to the new theories. Sixteen articles appeared in November alone.

Meanwhile, *The Times* (London) again impressed its readers with the importance of relativity in a leading article headed, "Modern Theory of Relativity." In a strong, clipped tone, the paper stated that "the ideals of Aristotle and Euclid and Newton which are the basis of all our present conceptions prove in fact not to correspond with what can be observed in the fabric of the universe . . . . Space is merely a relation between two sets of data, and an infinite number of times may coexist. Here and there, past and present, are relative, not absolute, and change according to the ordinates and coordinates selected."[7]

Einstein himself was besieged by reporters who came to Berlin from all around the world. In a letter to Max Born, he complained that the publicity was "so bad that I can hardly breathe, let alone get down to sensible work."[8] Einstein, nevertheless, cooperated with the reporters and also agreed to write an article on relativity for the London *Times* and give an interview-format article to *The New York Times*.[9]

Eddington explained relativity in public "Wonderland" lectures in which he discussed Alice's shrinking as an example of relativistic space. Reporting on his December 3 lecture, *The New York Times* said, "The greatest interest was shown in a lecture given last night at Trinity College, Cambridge, by Professor of Astronomy Eddington on Einstein's Theory, the dining hall being crowded with dons and undergraduates while hundreds were standing."[10] That a new physical theory should provoke such interest was news in itself. Many recognized that physics was generating concepts as revolutionary as those born in the Renaissance – perhaps more so.

A certain light-headedness, perhaps created by the difficulties in explaining the new theories, spawned a wide variety of cartoons and poems, especially such "relativity" limericks as the famous concoction of Arthur Buller:

> There was a young lady named Bright,
> Who traveled much faster than light.
> She started one day
> In the relative way,
> And returned on the previous night.[11]

The New York Times, in an editorial headed "Relativity at City Hall," remarked that "the rays of logic emanating from the Mayor's office are bent as badly as Einstein's rays . . . . And a man who annihilated space may be able to provide our municipal government with some happy thoughts on the rapid-transit problem."[12] *Punch* carried a cartoon showing police catching a burglar because the light rays from their flashlights bent around corners. Bertrand Russell and the mathematician Littlewood discussed the theory. "We used to debate whether the distance from us to the post-office was or was not the same as the distance from the post-office to us . . . . The eclipse expedition which confirmed Einstein's prediction as to the bending of light occurred during this time, and Littlewood got a telegram from Eddington telling him that the result was what Einstein said it should be."[13]

Einstein enthusiasts turned him into something of a popular star. The London Palladium asked him to appear for a three-week performance. He refused. More serious lecture invitations came literally from around the world. Many of these he accepted. *The Times Educational Supplement* (London) devoted three full-page articles to discussions of relativity by such respected experts as Lindemann and Whitehead. The House of Commons even started an Einstein Society in 1920.[14] The public eagerly paid for all the information it could get on the new theories.

Magazines, not to be outdone by the newspapers, joined the enthusiastic response.[15] In *The Contemporary Review* for December 1919, Eddington discussed the implications of his expedition's verification of relativity.

> Newspaper reports have been accurate on the whole; and it is evident that scientific writers of experience have tried nobly to cope with the editorial demand for an account of a most profound scientific theory, embracing all nature, 'in a simple language in not more than 500 words.' Perhaps the public has derived an impression that the new views of space and time are more weird than is really the case; but the fundamental nature of the change has not been exaggerated. The theoretical researches of Professor Albert Einstein, of Berlin, now so strikingly confirmed by the British eclipse expeditions, involve a broadening of our views of external nature, comparable with, or perhaps exceeding the advances associated with Copernicus, Newton, and Darwin.

Current Opinion for January 1920 carried an article headed, "The Most Sensational Event in Physics Since Newton: Father Time Goes Out Of The Picture." *The Living Age* for April 2, 1921, in an article taken from *The Manchester Guardian*, discussed the writer's enthusiastic response to an Einstein lecture which he found exhilarating and "perfectly clear." *The New Republic*, in an extensive two-part article early in 1920, explained the basics of the special and general theories and closed with the comment:

> Whatever may be the fate of the theory of relativity, it has undoubtedly opened up new regions of thought by suggesting new possible connections between fundamental ideas like energy, space, matter, gravity, etc., and can there be any greater service to the human mind than this opening up of new fields?[16]

A third article in *The New Republic* (July 6, 1921) gave reviews of eleven books on the new science and challenged "the irresponsible assertion that there are twelve men who understand" the theories. The claim here and elsewhere was that any layperson could, with some serious effort and study, understand the basics of the new ideas.

Not all magazines and journals were willing to make that effort, however. *The Nation*, in witty but ill-informed articles, tried at first to pass off relativity as something that psychologists, priests, and poets had known all along. It grouped relativity "in the same category with the annual sea serpent, the seven-year mutation of our bodies, the jargon of Freud, the messages from Mars . . . . Certain troubled spirits, hearing the law of gravitation called in question, do not feel sure that the earth may at any moment slip its Newtonian moorings and go ranging off out of gravitation into the ether – which we now hear does not exist."[17]

In a sarcastic comment *The Nation* claimed that "the difficulty of the subject is greatly exaggerated. If you do not understand it at a glance, you are a dunce and a simpleton."[18] Similarly, *The Atlantic*, in articles connecting the space-time world with the Mad Hatter's tea party in Wonderland, postulated the following:

> Everyone has read three or four popular explanations of this fascinating theory, and so is no doubt thoroughly familiar with its fundamental formulae; for the sake of historical background, however, I will give a brief but rigidly scientific résumé of the five principal points.
>
> A. Axiom: As we fly through the ether, the wind must blow in our face.
>
> B. Einstein's Idea: There is no way of finding out how the wind is blowing. From these it follows that, –
>
> C. All bodies are shortened in the direction of their motion.

D. The time of day depends on our direction and speed.
E. The distance from here to there depends on the wind.
(Note: As everyone knows who ever studied mathematics, the expression 'it follows' means 'if *you* don't you
flunk'.)[19]

Cartoons took similar note of the widespread interest in relativity,
as Fig. 1 demonstrates.

Even those journals which were originally sarcastic soon began
taking the whole matter more seriously. In June of 1925, *The Nation*
carried, in a series of four articles, Bertrand Russell's "ABC of
Relativity" before it was published in book form. Russell opened
with the remark, "It is generally recognized by the public that
Einstein has done something astonishing, has in some way revolu-
tionized our conception of the physical world."[20] In the same issues
carrying articles by Sherwood Anderson and Edwin Muir, and a
Joseph Wood Krutch review of Virginia Woolf's *Mrs. Dalloway*, one
could read:

> There is not, in objective nature, anything that can be called
> the time interval between two events. Different clocks give
> different estimates of the time, according to the way they
> have been moving, and there is no standard by which we can
> say that one of these clocks is right and the others are
> wrong.[21]

> Measurements of distances and times do not reveal properties
> of the things measured but relations of the things to the
> measurer. What observation can tell us about the physical
> world is therefore more abstract than we have hitherto
> believed . . . . Newton assumed absolute space and absolute
> time, and consequently absolute motion . . . . In a case of
> relative motion, [however] we cannot give any meaning to the
> statement that one of the bodies concerned is "really" at rest
> while the other is "really" in motion.[22]

In the final article Russell predicted that once we got used to
relativity, we would undergo important changes in our habits of
thought. We would deal with events rather than with things in
motion; we would work with greater abstraction; we would replace
the old laws and absolutes with relative truths and useful fictions.

The collapse of the notion of one all-embracing time, in

"People slowly accustomed themselves to the idea that the physical states of space itself were the final physical reality."
—PROFESSOR ALBERT EINSTEIN

Fig. 1. Drawing by Rea Irvin from *The New Yorker*, © 1929, 1957, by The New Yorker Magazine, Inc.; reproduced by permission.

which all events throughout the universe can be dated, must in the long run affect our views as to cause and effect, evolution, and many other matters. The poet speaks of

> One far-off divine event to which
> the whole creation moves.

But if the event is sufficiently far off, and parts of creation move sufficiently quickly, some parts will judge that the event has already happened, while others will judge that it is still in the future. This spoils the poetry. The second line ought to be:

> To which some parts of the
> creation move, while other parts
> move away from it.[23]

Naturally, science-oriented popular journals devoted considerable attention to relativity. In 1920, the *Scientific American* sponsored an essay contest for the best 3,000 word explanation of the Special and General Theories of Relativity. Editorials and commentaries, a $5,000 first prize, and rapidly growing interest brought a tremendous response to the contest. Writers from more than twenty countries participated, including such internationally known figures as Pickering, de Sitter, and Becquerel. The winner, in a delightful coincidence, was an obscure man working in the British Patent Office. (Einstein developed his Special Theory of Relativity while he worked in the Swiss Patent Office.) The *Scientific American* published the winning essay on February 5, 1921, and printed several other good essays from the competition in subsequent issues. A collection of the best was published in book form, under the title *Relativity and Gravitation*, in 1921.

Relativity, then, became suddenly fashionable, and Einstein found himself projected into the position of international culture hero. Fans smoked "Einstein cigars." *The Times*, which had done so much to spark the new fashion, began complaining in 1922 about the general simple-mindedness of the fad. "A sustained effort of the brain is necessary to penetrate the thought of the great German savant and to follow his logic. Thus the craze to discuss Einstein between two rubbers of bridge appears to be one of the funniest things in the world."[24] As Feynman points out, with some disdain, "When this idea descended upon the world, it caused a great stir among philosophers, particularly the 'cocktail party philosophers' who say, 'Oh, it is very simple. Einstein's theory says all is relative . . . . It has been demonstrated in physics that phenomena depend on your frame of reference'."[25]

To claim that everything depends on one's point of view, however, is so simple and unoriginal an idea that Einstein certainly did not have to go to all the bother of formulating the relativity theory in order to discover it. Indeed, as we shall discuss later, such a reduction misses the entire point of Einstein's relativity, which was to construct absolute relations, not demolish them.

Many physicists realized that the new theories were crying for some accurate and yet comprehensible presentation that would satisfy the intelligent layperson without offending the scientist. Hence, within five years of the eclipse expedition, scientists published a number of books popularizing the new theories – not only of relativity but also the new understanding of atomic phenomena, the wave-particle picture of reality, the quantum proposals, and the principles of uncertainty, wave mechanics, and complementarity as these developed. Einstein's own book, *Relativity: The Special and General Theory*, was translated from the German by Robert Lawson and was published in England (Methuen) and America (Holt) in 1920. Seven editions were turned out in nineteen months.[26] Eddington's *Space, Time and Gravitation* (Cambridge University Press), one of the most popular accounts in both England and the United States, appeared in 1920 also. Other works, often translated from the German or French originals, added to the publishing boom.[27]

In 1925, the two most prominent philosophers in England published their views on the new physics – Bertrand Russell in his *ABC of Relativity* (New York and London: Harper and Bros.) and A. N. Whitehead in his *Science and the Modern World* (New York and London: Macmillan). As Clark points out, "The relativity industry was flooding the continent, as well as Britain and the United States, with explanations that ranged from the erudite to the simpleminded."[28]

▶   *Einstein, the crowd-pleaser*

Einstein did much to popularize the new physics in his visits to America, England, and France. His shaggy and rumpled Chaplinesque appearance endeared him to photographers and cartoonists while his natural openness won his listeners. Reporters met him and the Zionist Chaim Weizmann as their boat landed in New York in April, 1921. The front page of the Sunday *New York Times* carried the headline, "Prof. Einstein Here, Explains Relativity; 'Poet in Science' Says It Is a Theory of Space and Time, But It Baffles Reporters." The opening of the long article described this unlikely hero:

> A man in a faded gray rain coat and a flopping black felt hat that nearly concealed the gray hair that straggled over his ears stood on the boat deck of the steamship *Rotterdam* yesterday, timidly facing a battery of cameramen. In one hand he clutched a shiny briar pipe and with the other clung to a precious violin. He looked like an artist – a musician. He was. But underneath his shaggy locks was a scientific mind whose deductions have staggered the ablest intellects of Europe.[29]

While in New York, Einstein lectured both at Columbia and at City College with reporters covering the lectures. Crowds greeted him everywhere. In New York, according to the *Literary Digest*:

> [The visitor] inspiring the most spontaneous popular demonstration is not a great general or statesman but a plain man of science – Dr. Albert Einstein . . . . It is something when New York turns out to honor a stranger bringing gifts of this recondite character. [As a reporter accurately predicted] the public will be glibly discussing the Einstein theory of relativity, whether or not it proves capable of understanding it.[30].

In Washington he met with President Harding, who admitted he was puzzled by relativity. Einstein was the guest speaker at the annual meeting of the National Academy of Sciences where he was introduced with the following tribute:

> The academy rejoices to bring its tribute of homage to the brilliant and penetrating mind which has so greatly enriched the philosophy of ultimate truth. We congratulate you on the universal appreciation of your investigations, which have outrun and overleaped the limitations and barriers associated with nationalities and with the times. To men everywhere your name, in association with the abstruse subject of your investigations, has become a household word.[31]

In Princeton, where he received an honorary degree, Einstein gave several lectures which were immediately published. These included discussions of the quantum theory, the wave-particle picture, and non-Euclidean geometries, in addition to the numerous elements of relativity theory. His American trip also included visits to Chicago and Cleveland, where the crowd caused a near riot. He was protected from possible injury "only by the strenuous efforts of a squad of Jewish war veterans who fought the people off in their mad efforts to see him . . . . Such a swirl of fighting, crowding humanity . . . has seldom been seen in Cleveland."[32]

In June 1921, Einstein arrived in England, going first to Manchester where he spoke on both Zionism and relativity. *The Manchester Guardian* commented on Einstein's popularity:

> The man in the street, a traveller between life and death, is compact of all elements, and is devoid neither of science nor of poetry. He may have few ideas in either, but he probably cherishes what he has, and whatever touches them nearly is of moment to him. Professor Einstein's theory of relativity, however vaguely he may comprehend it, disturbs fundamentally his basic conceptions of the universe and even of his own mind. It challenges somehow the absolute nature of his thought. The very idea that he can use his mind in a disinterested way is assailed by a conception which gives partiality to every perception.[33]

In London, Einstein stayed with Viscount Haldane, former Lord Chancellor, who knew Einstein's theories quite well and who spoke fluent German. Wisely protecting his guest, Haldane turned down a number of party invitations from fashionable London society. He did, however, hold an exceptional dinner party for Einstein. The guest list included the Archbishop of Canterbury (who had boned up on relativity beforehand with the aid of J. J. Thomson), Eddington, Dr. Inge (Dean of St. Paul's), Ian Hamilton, Harold Laski, Bernard Shaw, and Whitehead. As might be expected, wonderful and conflicting accounts of this dinner soon grew into myths.[34]

Einstein gave a well-received public lecture in London at King's College and spent a day at the Clarendon Laboratory at Oxford. From all accounts, his visit was a great success, not only for explaining the new physical theories but also for helping to heal the still fresh wounds of the First World War. *The Times* devoted ten articles to him during his two-week stay.

In France, where the war memories were even more painful,

Einstein's hosts carefully protected him from anti-German factions. He spoke in Paris to small, selected audiences which included Bergson and Marie Curie. Yet public interest ran high, and according to historian of science Michel Biezunski, Einstein's visit generated journalistic excitement beyond strictly scientific issues. "Was Einstein Swiss or German? And was or was not his theory in keeping with the French spirit? To these questions the newspapers carried contradictory answers."[35] Articles appeared in journals such as *L'Écho de Paris, L'Ère nouvell, Le Figaro, La France, Le Gaulois, L'Humanité, L'Internationale, L'Intransigeant, Le Journal du Peuple, La Lanterne, Le Matin, Le Populaire, Le Rappel,* and *Le Temps.* Charles Nordmann, an astronomer at the Paris Observatory and author of several excellent books and articles on Einstein, wrote two thorough accounts of the visit in *L'Producteur* and *Revue des deux mondes.*

Ronald Clark surveyed this period in his biography of Einstein: "The speed with which [Einstein's] fame spread across the world, down through the intellectual layers to the man in the street, the mixture of semireligious awe and near hysteria which his figure aroused, created a startling phenomenon which has never been fully explained."[36]

We shall suggest two major reasons for this spread of Einstein's fame and influence. First, his work had strong resonances with the contemporary revolutions occurring in art, music, theater, and literature. Through popularizations of his theories, Einstein served as a muse for these other revolutions. While his works were certainly not the only spurs to cultural innovation, Einstein proved to be a remarkably rich and authoritative source of idea and metaphor. Secondly, Einstein's personal image came to represent the power of scientific intellect. Einstein's face was a convenient literal image for discussions of genius, rational thought, and abstract notions. As the century wore on, the initial enthusiasms for scientific and technological progress turned to disillusionment with the mass mechanized deaths of two world wars, the failure of technology to banish poverty and injustice, and the creation of the ultimate threat of nuclear extinction. Einstein's service as the symbol of scientific intellect then acquired tragic overtones. Einstein became more than a muse. As we shall see, he came to be a mythic figure, representing both the light and dark aspects of science in society.

▶ *The parallel arts revolt*

As the world view of science rapidly shifted, artists were creating new expressions which corresponded to the new physical outlook. Picasso said, "Art does not evolve by itself, the ideas of people change and with them their mode of expression."[37] The majority of the public first became aware of dramatically different ways of dealing with time and space through the new painting rather than through the works of philosophers and physicists. Writers who were dissatisfied with traditional literary forms often drew from the major changes in art to institute parallel changes in their literature. Exposure to the new art also helped make the public more receptive to similar innovations in other fields, including physics. Those who championed the new arts were in touch with the avant-garde thinking in several fields, and Einstein's physics was a favorite comparison well before the general public became aware of Einstein. In 1913, for example, Apollinaire explained innovations in Cubism as follows:

> Until now the three dimensions of Euclid's geometry were sufficient to the restiveness felt by great artists . . . . Today, scientists no longer limit themselves to the three dimensions of Euclid. The painters have been led quite naturally, one might say by intuition, to preoccupy themselves with the new possibilities of spatial measurement which, in the language of the modern studios, are designated by the term: fourth dimension.[38]

Apollinaire, a poet and the major critic-champion of modern art before the First World War, was very much in the middle of the artists and musicians who were creating the revolution. So was the writer Maurice Raynal, a close friend of Picasso and his circle from 1904 on. In the Skira edition of *Modern Painting*, Raynal describes the atmosphere which helped generate the new art.

> Poetry, prose literature, philosophy and science – all were eagerly ransacked . . . . Hypothesis, assumption of a found solution, arguments *per absurdum* – all the procedures of the sciences and metaphysics – played their parts . . . in sponsoring the primacy of the thinking mind in the operations, no longer purely visual, of the creative process . . . . If, as I have said, science was a favored subject, this was not due to any special interest in new inventions – which I have heard

Picasso describe as "silly gadgets" – but rather to the emphasis laid on the mysterious operations of Chance, and all that these implied . . . . True, one heard talk about the fourth dimension, non-Euclidean geometry, the theory of numbers and so forth, not that any of us knew much about mathematics, but because these conceptions seemed to sponsor ventures on the artists' part into strange lands beyond the frontiers of conventional art, and to encourage creation rather than the "imitation" specified in Aristotle's famous, but (to us) obnoxious definition of art – the revolt against which in fact was largely responsible for the cubist revolution.[39]

In America the connection between the two revolutions was discussed by such critics as Thomas Jewell Craven, art editor of *The Dial*. In an article headed "Art and Relativity" Craven wrote:

Professor Einstein's revolutionary theory is the latest example of the eternal kinship between art and science . . . . While the celebrated physicist has been evolving his shocking theories of the courses of natural phenomena, the world of art has suffered an equivalent heterodoxy with respect to its expressive media . . . . The fixed coordinates upon which the Newtonian measurements were erected have their parallel in more than one aesthetic manifestation . . . and it is this abstract quality which establishes the analogy between the old art and classical mechanics. Professor Einstein's general theory of relativity has shaken the whole physical structure; similarly has the modern painter broken the classical traditions.[40]

According to Craven, the modern artist has done away with stasis, architectural symmetry, and perspective because he has been given a new "system of coordinates." Many creative people in avant-garde groups were aware of the changing world view and contributed in large measure to its expression. They kept in frequent communication with each other through travelling exhibitions, the exchange of little magazines, numerous visits (especially to the Paris group) and the international communication service provided by such men as Apollinaire, Roger Fry and Ezra Pound. Letters and journals indicate that discussions in these groups focused on certain subjects: the poetry of Mallarmé, the novels of Flaubert and Dostoevski, the art of Cézanne, the philosophy of Bergson and James, the psychology of Freud, and the science of the Curies, Planck, and Einstein. As Ortega y Gasset wrote, "The pattern of a new sensibility shapes itself before our eyes."[41]

Those trying to find direct evidence of the artists' awareness of the new physics often emphasize that the mathematician Princet lived in the *Bateau Lavoir* and smoked hashish with Picasso. Conflicting accounts make Princet either a brilliant mathematician with great influence on the painters or a mediocre accountant and a dope addict with no particular influence. With or without Princet, however, the revolution in art would have occurred for the reasons Raynal elaborated. Richardson writes, "The conditions for it were certainly 'in the air' . . . . Too, the immediate impact of the style [Cubism] on the painting of the avant garde, and even on its writing, is suggestive of some sort of inevitability. If one wished to find a parallel to Cubism in the sense of an inexorable force revolutionizing a field of intellectual endeavor, he will find the closest thing to it in modern physics."[42]

A far more likely source than Princet, if we insist on trying to find direct influences, lies in the painter Juan Gris. Gris studied university-level mathematics and physics before coming to Paris in 1906, when he "fell into the studio of Picasso."[43] Braque, Gris, and Picasso made up the great Cubist triumvirate, and Gris was by far the most articulate theorizer of the group. Many of his comments on the new art reflect his scientific orientation.

> The mathematics of picture making leads me to the physics of representation . . . . Only these mathematics are capable of establishing the composition of the picture . . . . Cubism is not a manner, but an aesthetic, and even a state of mind; it is therefore inevitably connected with every manifestation of contemporary thought. It is possible to invent a technique or a manner independently, but one cannot invent the whole complexity of a state of mind.[44]

Both science and art experienced parallel revolutions which shaped and expressed a new way of seeing. The art dealer Kahnweiler, a close friend as well as financial supporter of these artists, comments that "a radical change has taken place in human thought. It means that our *weltanschauung* has changed."[45] "Cubism is helping to disseminate this new metaphysical conception. The Renaissance spirit ended vulgarly in materialism which preached belief in an immutable and readily cognizable outer world; this was broken down by the Cubist painters as it was by the philosophers and physicists."[46]

Those people who strongly resented the new art were often those who had not kept current with the changing world view. Sigfried Giedion says, "The fact that modern painting bewilders the public is not strange: for a full century the public ignored all the developments which led up to it."[47] Ortega y Gasset pointed out in 1925 that such "people feel out of their depth and are at a loss what to make of the scene, the book, or the painting."[48] Critics attacking the wild new art were soon accompanied by those disparaging Einstein's wild physics.

Negative critics of the revolution in art were hostile and angry rather than indifferent. In her biography of Roger Fry, Virginia Woolf describes the reaction to the first Post-Impressionist exhibition in England. "The public in 1910 was thrown into paroxysms of rage and laughter. They went from Cézanne to Gauguin, and from Gauguin to Van Gogh, they went from Picasso to Signac, and from Derain to Friesz, and they were infuriated."[49]

In the United States, the famous 1913 Armory Show, which exhibited some 1,300 works by 300 European and American artists, caused an even more violent public reaction. "What the public saw was a new art which shocked rather than lulled. The 'fakers,' the 'madmen,' the 'degenerates' were abused, reviled and jeered. The press and the public laughed; the critics, with their standards crumbling around their ears, fulminated, but it was all to no avail. The greater the vituperation and hilarity, the greater the publicity and the greater the attendance."[50] Marcel Duchamp's *Nude Descending a Staircase* became the most talked about scandal of the show; it was called everything from an explosion in a shingle factory to a staircase descending a nude. Often the Cubists were treated as subversive anarchists or curious freaks. The pervasive atmosphere was one of a Barnum and Bailey side-show. The public exposure to the new art, however, caused a permanent shift in the attitude toward the moderns.

The first two decades of this century, then, witnessed astonishing changes in the way western culture understood reality. Both science and the arts (including literature) created revolutions in their approaches to time and space, and both revolutions were greeted with fascination, excitement, and often as not, outrage. The revolutions were related and contemporary, although the public became aware of Einstein's revolution only at the end of the period.

The traditional forms were unable to express the content of the new world view which modern physics helped to shape, just as the traditional Newtonian mechanics proved unable to explain the new world of the atom and the expanding universe. A brief summary of the Newtonian-Euclidean world view will help clarify the revolutionary nature of modern physics that replaced it, while an examination of the corresponding literature will demonstrate how pervasive, convincing, and successful Newtonian physics had been.

2

Newtonian mechanics and literary responses

Of Newton with his prism and silent face,
The marble index of a mind for ever
Voyaging through strange seas of Thought, alone.
Wordsworth

Time, space, and motion are fundamental terms that every civilization defines in attempting to describe the physical environment.
These terms may be simple descriptive measures, or may represent
elaborate philosophical formulations incorporating a religious
cosmology in which time and space come into existence by divine
creation.

Awareness of time comes from experiencing change. Motion, a
change in the place of an object with time, is thus a central means of
understanding time. A new representation of motion is central not
only to Duchamp's *Nude Descending a Staircase*, but to the scientific
revolutions of Newton and then of Einstein. By examining the
changing language of motion, time, and space, we can study
profound shifts in cultural world view.

Einstein's conception of motion, time, and space overthrew the
well-entrenched view of Isaac Newton, who had himself overthrown
the classical Greek formulation of Aristotle. Each of these philosophies of physical description were central components of broader
cultural world views, and indeed those cultural perspectives changed
dramatically as each revolution in science occurred.

▶ *The view before the renaissance*
Aristotle had begun his presentation of time, space, and
motion with the assumption that a state of rest is natural for
everything. All motion, therefore, must be caused by a force; when
the causal force stops, the motion stops. Aristotle's universe logically
depends on this initial assumption about rest – the Earth is at rest in
the center, and the planets and stars are moved or pushed in perfect

circles around it. The first pusher, or Prime Mover, was identified with God when Christianity absorbed Aristotelian concepts. In Aristotle's system a rock falls because its purpose or end is to be at rest in the sphere of its element, earth. Fire rises to reach its proper sphere.

Such fundamental assumptions form the base of a major world view which describes not only the physical world but also humanity's place in it, with implications about the meaning and purpose of life. Much of the world's great literature has been written in the context of this Aristotelian world view, including works by Dante, Spenser, Milton, and Shakespeare which reflect its hierarchical and teleological order. Shakespeare has his Ulysses make the analogy:

> The heavens themselves, the planets, and this centre
> Observe degree, priority, and place,
> Insisture, course, proportion, season, form,
> Office, and custom, in all line of order.
>
> . . . But when the planets
> In evil mixture to disorder wander,
> What plagues, and what portents, what mutiny,
> What raging of the sea, shaking of earth . . .
> Divert and crack, rend and deracinate
> The unity and married calm of states
> Quite from their fixure! O, when degree is shaked,
> Which is the ladder of all high designs,
> The enterprise is sick![1]

The static social order parallels the static physical order. Just as Aristotle had seen each physical object as having an appropriate place in space, so in the larger world view everything had its place along great chains of being, fixing social, economic, biological, and religious hierarchies as well as the physical universe.[2]

► *The first revolution*

The Newtonian revolt against Aristotle's world view was an amazingly practical success. In later hands Newton's scheme was able to predict a new planet (Neptune) and facilitate the Industrial Revolution. A Newtonian world view came to dominate western thinking for over two hundred years. Copernicus, Kepler, Galileo, Newton, and others accomplished this revolution by providing different and increasingly comprehensive answers to the same basic questions about time and space, and about the structures, operations, and events in the universe. Instead of Aristotle's doctrine of rest, for example, the newer physics provided the concept of inertia – according to Newton's first law, a body will continue either at rest *or* in motion at a constant velocity, in a straight line, unless a net force acts on it. Hence the stars and planets needed no continuing pushers or Prime Movers to have straight-line motion. Curved motion or changes in velocity were the result of forces between bodies. Given the strength and direction of a force on an object, and the mass of the object, Newton's second law provides a formula for the change that the force will cause in the motion of the object.

Gravity is one dominant force, and Newton also produced a formula for its strength and direction. The differing descriptions of the nature of gravity for Aristotle and Newton clearly show the distinction between their views of science. For Aristotle, each earth-like object was endowed with a disposition to fall. That tendency remained regardless of environment. With Newton, inter-action between bodies became the basis of all forces, including gravity. Robert Karplus identifies the substitution of interaction for predisposition as the major change between the old science and the new. Since the Newtonian revolution, "the scientist ascribes happen-ings to interactions among two or more objects rather than to something internal to any one object. Thus, the falling of an apple is ascribed to its gravitational interaction with the earth and not to the heaviness inherent in the apple."[3] With this picture comets falling to the sun or moons orbiting Jupiter could be understood as natural phenomena, and not as improper bodies wandering to disorder.

The motions of falling apples and orbiting planets thus depends on linkage with other bodies. The linking force is Newton's universal gravity. The future motions of any object can be predicted from knowledge of their initial positions and velocities, and knowledge of the laws of just the few universal interactive forces.

Aristotle's and Newton's methods did both produce similar descriptions for some aspects of the world, particularly those that were treated accurately by the older system. Aristotle would say that a rolling stone stops when it is no longer forced to move, and can return to its preferred state of rest. Newton would also predict the stone would stop, but due to frictional forces – interaction between the stone and the ground – that change the stone's motion. Newton's explanation is not immediately superior, although with further development Newtonian mechanics was able to predict how long it would take for the stone to stop, a quantitative result beyond the power of the older scheme.

Newton's more dramatic successes were in astronomy. When Newton related his laws of motion to Kepler's earlier laws of planetary motion, he was able to test, beyond the Earth, his law of universal gravitation – certainly one of the most astonishingly simple and successful laws of all time.[4] With this law scientists successfully predicted the existence and position of a previously unseen planet, Neptune. The law also showed the connection between a falling apple and such seemingly independent phenomena as the orbit of the moon and the pattern of the tides.

By mid-19th century, Newton's laws enabled celestial mathematicians to draw up a table of eclipses covering future centuries, and accurate to a matter of seconds. Comets, feared as evil omens from the beginning of history, could now be identified by Newtonian calculations as sun-orbiting celestial balls of gas and dust making regular rendezvous with the Earth. The more universal predictive abilities of Newton's science, especially in the realm of planetary motions, soon convinced scientists of the superiority of the new methods.

Outside of the small community of mathematicians and natural scientists, however, the shift from Aristotle to Newton took longer and was more difficult, because Aristotelian science was only part of a much larger world view of ideal order. If it was wrong to believe that objects followed divine purpose in seeking rest, then was it also wrong to believe in the divine right of kings to command social rest? In Newtonian mechanics, physical objects have no proper place or state of motion, but merely respond to other objects in the environment through universal forces such as gravity. Contrary to Shakespeare's planets, each of which had a proper orbit, each of Newton's planets could occupy any of an infinite selection of orbits and could even be deflected, by a gravitational tug from a passing celestial body for example, to take up a new orbit. Only the configuration of objects and forces at any given moment determined behavior. Science no longer supported the great chains of being.

The immediate effect of Newton's cosmology thus seemed to those outside the scientific community to be the destruction of a scheme which had provided a unifying order. With the physical universe no longer providing a model of propriety, the chaos Ulysses feared in

Troilus and Cressida would be the eternal state of both heaven and earth. John Donne expressed the apprehension of his contemporaries as they grappled with the "new Philosophy":

> And new philosophy calls all in doubt,
> The element of fire is quite put out;
> The Sun is lost, and th'earth, and no man's wit
> Can well direct him where to look for it.
> And freely men confess that this world's spent,
> When in the planets, and the firmament
> They seek so many new; then see that this
> Is crumbled out again to his atomies.
> 'Tis all in pieces, all coherence gone;
> All just supply, and all relation:
> Prince, subject, father, son, are things forgot,
> For every man alone thinks he hath got
> To be a phoenix, and that then can be
> None of that kind, of which he is, but he.[5]

▶   *The universe as clockwork*

But if Newton's synthesis removed support from one general world view, it provided props for a new outlook. In a Newtonian universe the determination of future events could not be made from principles of propriety, as it could have been under Aristotle's vision. However, with knowledge of the linking mechanical force laws and of the locations and speeds of all bodies at any one moment, such a determination seemed feasible again. Once set in motion, the physical universe from atoms to living cells to stars would simply respond to the net effect of inertia and interactive forces. In theory all future behavior would be determined, much as the motion of the hands of a clock could be predicted, given the configuration of gears at any one moment and the force of the spring.

If the physical universe were no more than a fantastically complex clock, then all physical events would be completely determined. God no longer needed to be continuously involved with moving the planets – once set in motion, they would continue forever their mechanical destinies. The magnificent clocks and automata of Germany in the century just before Newton's *Principia* could reproduce in miniature the motions of the planets and stars, and could even mimic simple animal and human gestures. After the

Newtonian view had pervaded the whole society, these clocks could be seen not just as mimics of the external behavior of the universe, but as models revealing its internal principles. A new metaphor, the clockwork universe, was born from the new science.[6]

The successes of Newton's theory in astronomy created a widespread belief that eventually all aspects of the universe, even social relations, would be explained by analogous mechanical models. This belief was encouraged by a reasonable desire to explain the unfamiliar in terms of the familiar. Twentieth-century historian of science d'Abro has noted: "The possibility of devising a familiar model of underlying physical occurrences was long regarded in England as an argument in favor of a theory. [The nineteenth-century physicist Lord] Kelvin once said he could not understand a phenomenon unless he could devise a mechanical model of it."[7]

▶   *The poets' fascination with Newton*
Not everyone was delighted with this clockwork mechanism, as Arthur Clough's poem indicates:

> And as of old from Sinai's top
>   God said that God is one,
> By Science strict so speaks He now
>   To tell us, There is None!
> Earth goes by chemic forces; Heaven's
>   A Mecanique Celeste!
> And heart and mind of human kind
>   A watch-work as the rest![8]

Both Marjorie Hope Nicolson and Douglas Bush have published excellent critical studies on the effect of science on poetry, especially the impact of Newtonian mechanics and optics and of the Copernican-Galilean cosmology verified by Kepler's and Newton's laws.[9] Nicolson traces the early enthusiastic response to the new physics which, in England, came close to making Newton a demigod. In particular, the public was fascinated with his *Opticks*.

In that volume he analyzed white light and demonstrated for the first time that white light is a mixture of colors in the spectrum, and that a glass prism physically separates the colors by refracting blue light more than red. The colors once separated are pure – not mixtures themselves. Newton had thus provided mechanical laws for both tangible matter and intangible light, although his inability to merge the two sets of laws could have been a warning of difficulties to come.

"There is no question," Nicolson writes, "that the first effect of Newton's resolution of the colors and his careful analyses of their properties was to produce a new scientific grasp of a richer world of objective phenomena peculiarly sympathetic to poets."[10] James Thomson, in his "To the Memory of Newton," exemplified the poetic response to the *Opticks*:

> Even light itself, which every thing displays,
> Shone undiscovered, till his brighter mind
> Untwisted all the shining robe of day;
> And, from the whitening undistinguished blaze,
> Collecting every ray into his kind,
> To the charmed eye educed the gorgeous train
> Of parent colors...[11]

White light from the sun, separated into spectral colors by rainbows, gems, or insect wings, provided basic imagery and symbolism for much 18th-century poetry. So did an admiration for those reasoning powers which enabled humankind to discover nature's laws. John Reynolds, Richard Blackmore, Henry Brooke, and other "scientific poets" analyzed problems of the physics of light and discussed Newton's theory that light was corpuscular, that it travelled in straight lines, that it was refracted by a change in medium. As Nicolson points out, these scientific poets, "while they instructed, could hardly be said to delight."[12]

Perhaps the least poetic of subjects, the mechanical laws mathematically described in Newton's *Principia*, still found their way into poetry. According to Bush, "In general the carrying of a scientific ideal into poetry favored direct logical statement as against oblique suggestion and symbol."[13] James Thomson was more successful than most in transforming Newton's law of gravity into poetry:

> . . .by the blended power
> Of gravitation and projection, saw
> The whole in silent harmony revolve . . .
> And ruled unerring by that single power
> Which draws the stone projected to the ground.[14]

Science has limitations as a guide for poets or as a model for human behavior. As these limits became clearer to later writers, the celebrations of the magnificent clockwork universe gradually changed into an artistic revolt against rationalism and materialism. Alexander Pope complained about the undue arrogance of Newtonians who applied physics to many non-scientific disciplines as if it were the one and only eternal truth.

> Go, wond'rous creature! mount where Science guides
> Go, measure earth, weigh air, and state the tides;
> Instruct the planets in what orbs to run,
> Correct old Time, and regulate the Sun . . .
> Go, teach Eternal Wisdom how to rule –
> Then drop into thyself, and be a fool![15]

Pope's satire belongs to the Age of Reason, and it was a rational protest against unbalanced reliance on mechanics at the expense of religion and common sense.

> We nobly take the high Priori Road,
> And reason downward till we doubt of God . . .
> Thrust some Mechanic Cause into his place;
> Or bind in Matter, or diffuse in Space.
> Or, at one bound, o'erleaping all his laws,
> Make God Man's Image, Man the final Cause . . .[16]

Wisdom, Pope claims, lies in recognizing that God is "the source of Newton's light, of Bacon's Sense."[17]

Many of the Romantics, however, went much further than Pope, rejecting Newton's physics altogether. Blake designated Newton as a devil and science as the Tree of Death. For Blake, scientific empiricism and mechanics "mock Inspiration and Vision," which are the poet's elements. Keats in the famous passage from *Lamia*, cries out against the heartless mechanics of Newton's *Opticks*:

> Do not all charms fly
> At the mere touch of cold philosophy?
> There was an awful rainbow once in heaven:
> We know her woof, her texture; she is given in the
> Dull catalogue of common things.
> Philosophy will clip an Angel's wings,
> Conquer all mysteries by rule and line,
> Empty the haunted air, and gnomed mine –
> Unweave a rainbow . . .[18]

Coleridge, in "The Theory of Life," argues that from the time of

Kepler and Newton "not only all things in external nature, but the subtlest mysteries of life and organization, and even of the intellect and moral being, were conjured within the magic circle of mathematical formulae."[19] The Romantic poets, trusting in imagination and intuition, rebelled against both the technological extension of mechanics into industrialized society and the philosophical limitations of rationalism. Thus Blake protests against the drudgery and dehumanization of "these dark Satanic mills," as well as against the narrowness of "single vision" – reason alone – without intuition, inspiration, imagination, and feeling. A world view limited to materialistic science, then, is "under-dimensioned"[20] because it eliminates the living observer as a dynamic participant.

Alfred North Whitehead recognized that Wordsworth and other Romantics had not only realized the limits of materialism but had proposed an organic philosophy which anticipated aspects of modern science – especially in its concerns with change, value, and involvement of the observer with the observed. Coleridge denied Descartes' division of nature and intelligence, finding them organically related in an ongoing process. For Coleridge, a poem should be an organic form, like a tree, shaped from within, as opposed to a mechanical form which is imposed on the material from the outside. The Imagination provides "the very powers of growth and production."[21] Reality, then, becomes an organic whole which art can express better than a reductive, analytic science.

America saw a similar Romantic rejection of materialism and mechanism. Melville, for instance, in his story "The Tartarus of Maids," transforms the paper mill into a virgin-destroying fiend, a mechanized devil. Poe, in his famous "Sonnet – to Science" compares science to a vulture that preys upon the poet's heart and that drives out all myth and dream. Both Emerson and Thoreau anticipated the modern process philosophies with their vision of the natural world as a rushing stream or dynamic organism which will not stop to be precisely observed and measured. Nevertheless, as Douglas Bush points out, Americans have generally accepted science and technology as assets. "New world optimism cushioned Emerson, Thoreau, and Whitman against the fear of science."[22] Thus, both Whitman and Emily Dickinson could celebrate the locomotive:

> Thy ponderous side-bars, parallel and
> connecting rods, gyrating, shuttling at thy sides,

> Thy metrical, now swelling pant and roar,
>   now tapering in the distance . . .
> Type of the modern – emblem of motion and power –
>   pulse of the continent . . .

(from Whitman, "To a Locomotive in Winter"[23])

> I like to see it lap the Miles –
> And lick the Valleys up –
> And stop to feed itself at Tanks –
> And neigh like Boanerges –
> Then - punctual as a Star
> Stop – docile and omnipotent
> At its own stable door

(from Dickinson, "I Like to See It Lap the Miles"[24])

This romantic animation and affectionate treatment of the machine soon changed. Naturalists portrayed machinery as dehumanizing and destroying life (Frank Norris in *The Octopus*, Dreiser in *An American Tragedy*, Steinbeck in *The Grapes of Wrath*, O'Neill in *The Hairy Ape*, and others). T. S. Eliot described the modern world as a Waste Land, choked with industrial pollution and emptied of meaning and vitality by the indifferent mechanisms of society.

▶ *Determinism as a world view*

Philosophically, Newtonian mechanics had implied a strict cause-effect determinism for material bodies, which seemed to many to deny any free will or creative originality for humanity. The pervasive pessimism of literary naturalism derives partly from this mechanical determinism. For some writers, however, determinism was not an enemy to be attacked but rather a condition to be worked with in writing. George Becker describes the major aspects of literary realism as reflections of the Newtonian causality. "The key to the realist position," he writes, "is that the universe is observably subject to physical causality; man as part of the physical continuum is also subject to its laws, and any theory which asserts otherwise is wishful thinking."[25] What the realist reflects, when he functions as a mirror walking down the road, is the visible and tangible environment subject to mechanical determinism. Of course, none of the great realistic novelists was so simply limited.

Zola, in his influential essay, "The Experimental Novel," writes that "there is an absolute determinism for all human phenomena." Therefore, novelists should apply the scientific method to writing. "Determinism governs everything. It is scientific investigation; it is experimental reasoning that combats one by one the hypotheses of the idealists and will replace novels of pure imagination by novels of observation and experiment."[26] The author, then, becomes both an observer and experimenter who "sets the characters of a particular story in motion, in order to show that the series of events therein will be those demanded by the determinism of the phenomena under study."[27] For Zola, only by recording observable reality and by experimenting with variables could any knowledge of reality be gained, either in the laboratory or in the novel.

Another assumption of Newtonian mechanics involved a separation of subject and object. The observer could be completely objective; his observation in no way interfered with what was going on. Zola wrote, "Like the scientist, the naturalist novelist never intervenes."[28] Flaubert put the same concept as, "One ought not let his personality intrude. I believe that Great Art is scientific and impersonal." Art should "rise above personal feelings and nervous susceptibilities. It is time to give it the precision of the physical sciences, by means of a pitiless method."[29]

Many writers, however, found determinism to be an outrageous bondage and the scientific method to be too limited a way of knowing. Experimental procedures could not account for truths known through faith, intuition, or imagination. In other words, the romantic objections to Newtonian mechanics kept resurfacing, long after the Romantic movement itself gave way to realism and naturalism. Stephen Crane, in "War is Kind," portrays a world whose laws remove the function of a personal supreme being who might have some compassion for man.

> A man said to the universe:
> "Sir, I exist!"
> "However," replied the universe,
> "The fact has not created in me
> A sense of obligation."[30]

And Walt Whitman, who could warm up to a locomotive, found himself distressed at the scientific explanations of the heavens.

> When I heard the learn'd astronomer,
> When the proofs, the figures, were ranged

in columns before me,
When I was shown the charts and diagrams
    to add, divide, and measure them,
When I sitting heard the astronomer where he
    lectured with much applause in the
    lecture-room,
How soon unaccountable I became tired and sick,
Till rising and gliding out I wander'd off
    by myself,
In the mystical moist night-air, and from
    time to time,
Look'd up in perfect silence at the stars.[31]

The line lengthens and lengthens as the poet's discomfort increases until the release comes and he is free to respond with his senses and emotions to the stars. The poem, then, expresses very well the artistic necessity to get beyond mechanics.

The major creative writers, of course, did not work in pigeonholes labelled Realism, Naturalism, Romanticism, Symbolism or whatever. Even Zola could not adhere to his own pronouncements about how writers should work; and his novels are the better for not being perfectly "experimental." Nevertheless, the impact of the Newtonian world view on everyone's daily life was inescapable, and it inevitably provoked both hospitable and hostile reactions in literature. The solidity of the mechanical scheme for solving problems in physical science was so clearly evident, however, that the press to accommodate a Newtonian world view grew stronger and stronger.

▶   *Broader impacts*

By the end of the 19th century, physics and its philosophical implications (no matter how distorted) had influenced many aspects of western thinking. Philosophers, sociologists, economists, historians, even artists framed "laws" that "determined" or at least described the determination of their respective subjects. G. H. A. Cole, in an essay in *The Twentieth Century Mind*, summarized this influence:

> The arguments of classical physics have been applied by some writers in the past to the wider content of human behavior to suggest that the full behavior of a person could, in principle at least, be reduced to such a calculation [future

behavior determined by present state]. More than this, the idea has been applied to the whole evolution of the world from which the conclusion is snatched that, according to the classical arguments of physics, the whole evolution of the universe was determined by God in the very process of creation. While this may be comforting to the sinner who is happy as he is, this whole argument proved embarrassing to religious apologists who insisted on the existence and value of free will.[32]

For some, then, God was the master clockmaker who had set the whole mechanism running, and humans were just cogs in the wheel who occasionally suffered under the delusion that they had freedom to change things.

Sociologists and economists applied scientific methods and postulated "scientific laws" for their own disciplines, and in so doing seemed to give their theories scientific sanction and credibility. Marx and Engels, for example, established "unconditionally valid rules; applicable not only to social sciences but to natural and physical sciences as well." Engels' "ramblings in science were elevated into a canonical text whereby party philosophers tried to decipher what course science ought to follow.... Marxist dialectic, like Comtean positivism, was not hesitant about laying down a long array of regulations physics was supposed to comply with."[33] Lenin claimed, "Modern physics is in travail; it is giving birth to dialectical materialism."[34]

Herbert Spencer derived his social Darwinism both from classical deterministic arguments of physics and the theory of evolution which seemed to fit right in with the cause-effect picture. Animals, or people, survived if their biological makeup was in tune with environmental conditions (food supplies, population densities, etc.). Economic "scientists" established laws for population densities in relation to food supplies and to supply and demand. Zola, who set up the rules for the experimental novel, found the determining causes of human behavior in both inherited and environmental factors. The author was to be the objective observer recording the data for his experimental novel. The whole naturalistic school bloomed in this heyday of determinism.

▶   *Seeking the limits to Newton's powers*
As successful as Newtonian physics was, almost all of the

real events of the physical universe were too complex for it to handle. Predicting the orbit of a planet around the sun is about as simple a problem as exists for physics, since only two bodies are involved, and only one significant force, gravity. Even so, a standard treatment would begin by simplifying further, and reducing the real planet-sun system to an ideal *model*, a universe of only two point masses and a single force, Newton's universal gravity. The predicted motion of these model points would then be compared with the real planet's motion about the sun to see how well the model, and the physics, were doing.

If we try to calculate the orbits of *three* mutually interacting celestial bodies, the problem becomes so complex that exact solution is not possible, although mathematical approximations can eventually yield predictions of the orbits to any desired precision. Beyond three interacting bodies the difficulties mount rapidly. A baseball thrown in the air collides with trillions of molecules every second. Air resistance and friction are forces far too complex for exact Newtonian calculations, although again clever approximations may be very helpful. In the end, however, practical problems such as the path of a baseball or a rolling stone always involve empirical measurements as essential steps if Newtonian physics is to make any useful predictions at all. Except for a very few simple situations, all of the wonders of Newtonian mechanism were more useful when applied to theoretical models than to precise descriptions of the real world inhabited by people.

As long as physics always proved correct within the known inaccuracies and uncertainties of the approximations, there was no reason to think that any of the physics was wrong. Scientists recognized that there were entire areas that had not been studied, such as the behavior of bodies of huge mass (the Sun or bigger), tiny mass (atoms or smaller) or high speeds (objects moving faster than sound). Still, two centuries without a major instance of nature clearly violating Newtonian models gave physicists reasons to believe that they finally had discovered the true laws of the universe. As long as physicists succeeded in applying Newton's principles successfully to every problem they did tackle, Newton would reign supreme.

Well into the 19th century, nearly all of those problems in mechanics that could be treated were treated successfully. Only a few cases failed outright. Thermal properties of matter were in general, but not perfect, agreement with theory. In the celestial realm, where Newtonian models were first successful, increasing accuracy of measurement continued to verify Newton's laws, with one significant exception: Mercury, the closest to the sun and swiftest of all the planets, orbited a tiny bit differently than predicted. As the astronomers made more refined calculations, they could take more and more small effects into account in the model, such as adding the effect of the gravitational tug of Venus on Mercury, but they were always left with a persistent difference between Mercury's predicted and actual paths. These nagging minor difficulties with mechanics were only exacerbated by the increasing scope and precision of experiment.

▶   *The problem of light*
     After two centuries of Newtonian physics, one major part of the human environment had still not been reduced, even crudely, to any quantitative mechanical model. That phenomenon was light – ironically, the subject in which Newton achieved his earliest fame. Newton's *Opticks* had revealed many new characteristics of light, and gave formulae for predicting light's behavior in some circumstances. We have seen how poets celebrated that success. But Newton had not succeeded in deriving his formulae for light from his principles of force and motion, and he had only guesses about the nature of light itself.

     What was light? Did it have the common physical qualities of mass, inertia, or weight? We are enveloped in it, depend on it to function, but it is literally intangible. Newton believed that light consisted of tiny particles, for which the same laws of force and motion that worked for planets *should* have worked, but Newton had not constructed a quantitative theory he could test. A convincing argument that light was *not* a stream of particles was provided by Thomas Young in 1801. He demonstrated a behavior of light that could not occur in a stream of Newtonian particles.

     Young demonstrated that intersecting beams of light could somehow interfere with each other, cancelling each other out in certain places and producing dark bands. That behavior is one of a class widely known today, called "interference" phenomena. One simple example of interference can easily be produced. Look with one eye at a bright, uniform source of light such as a blue sky or a white wall. Hold up one hand, very close to your eye, and look through the gap between two fingers. If you squeeze your fingers together until the gap is nearly closed, you will see dark bands parallel to your fingers appear in the gap.

The example using fingers to provide a single slit in which interference occurs is today called a "diffraction" effect, and is somewhat complex to describe quantitatively. Young set up a conceptually simpler experiment in which light passed through two slits, after which the two beams of light intersected. Where they crossed, they produced clearly visible dark bands. With this elegant and easily measured example, Young was able to show that the positions of the dark areas were precisely where they would be expected to be if the light passing through each slit was in the form of a wave. Any time that the crest of a wave from one slit arrived at the same place as the trough of a wave from the other slit, the two waves would cancel each other out, resulting in a dark space. Young's formulae could predict the location and spacing of dark and bright areas as a function of the slit size and color of light used.

Precisely analogous phenomena were studied with waves in water, where cancellation of the waves from two or more sources resulted in calm water at certain locations in a harbor (equivalent to the dark areas in the light experiment). For sound, waves in air interfering explained the bothersome dead spots at certain seats in concert halls.

In Young's widely repeated interference experiments, light behaved just like an extremely short wave, although there was no apparent medium, comparable to the water for water waves, or the air for sound waves, to carry the waves of light. Nevertheless, Newton's model of light as small particles could not even begin to predict these interference phenomena. How could particles cancel each other out?

Physicists were not too upset by this development since they knew that waves, as well as particles, could be good Newtonian models. Newton's mechanics were devised to explain simple particle motion, and worked wonderfully. Waves of water or sound were collective phenomena of huge numbers of interacting particles (comprising a medium such as water molecules or air molecules), and the physics of waves had to be just another instance of the need for empirical formulae when the situation was too complex for direct applications of Newton's equations for each individual particle. So if light were not a beam of individual particles, light could still be brought into the mechanical fold as soon as the nature of the waves and the nature of their medium were determined.

▶   *A second Newton emerges*
Finally, in 1873, the specific identity of the waves called light seemed to be revealed with the conclusion of James Clerk Maxwell's theory of electricity and magnetism. For several centuries, scientists had been compiling data about these two categories of phenomena. Equations had been found to describe the forces of attraction or repulsion between two magnets, and between two bodies charged with static electricity. Then connections between magnetism and electricity had been studied. A magnet moving near a coil of wire could induce an electric current to flow in the wire. Conversely, a strong current forced through a wire (for example, by a chemical battery) would produce a magnetic field around the wire, a field which could be seen by the deflection of a compass needle placed near the wire. Equations had been found for many different electrical and magnetic phenomena, and similarities among some of the equations had been noted.

In an achievement often compared to Newton's, Maxwell combined all known laws about electricity and magnetism into a set of four short equations. These established the nature of electric and magnetic forces as clearly as Newton's equation of gravitation had brought that force into the realm of the precisely calculable.

But Maxwell also added a term to one of his four equations, a factor from his own imagination. The resultant equation described an additional relation between electricity and magnetism that had not been observed experimentally. The particular relation by itself was not of great impact, but taken with the other three equations, the set took on an enormously increased symmetry and importance.

With Maxwell's added term, the equations predicted that a changing electric field could produce a magnetic field, and in turn a changing magnetic field could produce another electric field. The perfect reciprocity implied that a self-perpetuating cycle could be started. A changing electric field, begun by vibrating electric charges or some other method, would set up a changing magnetic field, which would set up a changing electric field, *ad infinitum*. Once begun, the alternating fields of electricity and magnetism would propagate, spreading out into space. A "wave" of electromagnetic energy would be produced.

Might these theoretical waves be realized in our world? Could light in fact be one of these electromagnetic waves? Maxwell calculated the speed at which his electromagnetic waves would spread through space. He used only his equations, and numbers taken from measurements of electrical and magnetic properties in conventional laboratory experiments. He was careful to note that "The only use made of light in the experiment was to see the instruments."[35] Remarkably, the speed of Maxwell's theoretical electromagnetic waves was just equal to the independently measured speed of light itself.

A great puzzle, the nature of light, seemed solved at last. Light was a wave of electricity and magnetism. The larger the amplitude of waves, the brighter the light. Color could be related to the frequency of the wave, how rapidly the fields rose and fell. The faster the frequency of the wave, the bluer the light. Red light was a wave of slower frequency than blue light. And new kinds of light could be found – colors invisible to the eye caused by frequencies slower than red, or faster than blue. Infrared and ultraviolet light were now part of a continuum, which eventually embraced radio and radar at the low frequency end, and X-rays and cosmic rays at the higher end. A grand unification had been achieved, and combined with Newton's equation for particle motion, all phenomena again seemed within the grasp of physics.

▶   *The program for physicists after Maxwell*
The last quarter of the 19th century might have been the time of polishing the final glorious structure of physics. The entire universe was made up of particles of matter which interacted through forces. Those forces were two: gravity, described by one of Newton's laws and acting between all particles, and electricity and magnetism, described by Maxwell's laws and acting on electrically, magnetically charged particles. The behavior of particles subjected to any forces could be predicted by Newton's equations of motion. Great collections of particles could interact, as a medium, and the resultant behavior could successfully be described as waves.

But instead of completing the structure of physics, Maxwell's triumph had established a new foundation, one that did not properly square with Newton's. The apparently seamless description of the universe in the preceding paragraph omits some critical problems. What was the medium in which light, a wave of electricity and magnetism, propagated? Light could travel through the vacuum of space, so what invisible electric and magnetic medium was waving there? What caused the particular assortments of waves (the colors) given off by various glowing sources, from stars to burning embers? Serious practical and then increasingly fundamental difficulties emerged as physicists struggled to combine Maxwell's successes with Newton's, while the small catalog of failures of Newtonian mechanics continued to muddy the situation. Eventually two separate revolutions would be required to straighten out the structure of physics: relativity and quantum theory.

Thus even as the Newtonian world view was spreading throughout every part of culture, scientists were becoming increasingly distressed that physics was diverging into two poorly connected structures, Newton's and Maxwell's, instead of converging into an ever simpler, all encompassing concept. By the end of the 19th century, theological poets like Francis Thompson, undaunted by these difficulties in recent physics, celebrated the reconciliation of religion with the absolute determinism and universal gravitational linkages of Isaac Newton:

> All things . . . linked are,
> That thou canst not stir a flower,
> Without troubling of a star.[36]

In those same years, Albert Einstein began his study of physics, and two new revolutions were already under way.

3

Einstein's revolution

Gentlemen! The ideas on space and time which I wish to develop before you grew from the soil of experimental physics. Therein lies their strength. Their tendency is radical. From now on, space by itself and time by itself must sink into shadows, while only a union of the two preserves independence.

Herman Minkowski, 1908

The physics young Albert Einstein studied in the 1890's was a beautiful, powerful set of concepts that could successfully explain a host of phenomena throughout the known universe. Western civilization had been led by that physics to discard its earlier metaphor of a great chain of propriety in favor of a metaphor of multiple, causal linkages, the great world-machine clockwork.

But physics in the last decade of the 19th century was increasingly troubled by a few phenomena that violated Newton's principles, such as the failure of Mercury to follow precisely its predicted orbit. Other profound puzzles lay in the failure of Newtonian physics to merge with the physics of light.

Maxwell had shown that light could be described as a wave of electricity and magnetism. Everyone was already familiar with waves of water, and just as the shape of a wave could pass across the ocean, so too could a wave of magnetism pass through an iron bar as a magnet was brought near, or a wave of electricity pass along the electric charges in a metal wire. These wave disturbances were carried by some identified medium: water, magnetic particles, or electric charges. But light could travel through the apparent vacuum of space. The regions between the planets and the stars seemed empty of any medium to carry the electric or magnetic waves, yet we *see* the sun and stars. Thus an outstanding fundamental question to be solved by Einstein's generation of physicists was: what was waving as light waves travelled through "empty" space? Could there be water waves without the water? That would be like Lewis Carroll's Cheshire Cat: the cat vanishes, but somehow the grin remains.

▶   *The rebirth of the ether*
In school, young Einstein learned that since we could see the

sun and stars, the space between the planets and stars had to be "filled with an extremely fine, imponderable substance, the ether, which is the carrier or medium of these phenomena."[1] The ether concept went back to Aristotle's perfect ether – that quintessence which composed the heavens. Electric and magnetic forces were believed to be transmitted by the ether as continuous forces or as oscillating waves called light. The "luminiferous" (light-bearing) ether was a substance that should also obey Newton's laws of mechanics, and it seemed for a few years the ether might unite Newton's mechanics with Maxwell's waves.

Maxwell was delighted with the idea of the ether carrying his electromagnetic waves across interstellar space.

> The vast interplanetary and interstellar regions will no longer be regarded as waste places in the universe, which the Creator has not seen fit to fill with the symbols of the manifold order of His kingdom. We shall find them to be already full of this wonderful medium; so full, that no human power can remove it from the smallest portion of space, or produce the slightest flaw in its infinite continuity. It extends unbroken, from star to star; and when a molecule of hydrogen vibrates in the dog-star, the medium receives the impulses of these vibrations; and after carrying them in its immense bosom for three years, delivers them in due course, regular order, and full tale into the spectroscope of Mr Huggins, at Tulse Hill.[2]

Helmholtz said, "There can no longer be any doubt that light waves consist of electric vibrations in the all pervading ether," and Poincaré claimed, "The ether is all but in our grasp."[3] FitzGerald, in an enthusiastic address to the British Association, stated, "It is only within the last few years that man has won the battle lost by giants of old, has snatched the thunderbolt from Jove himself, and enslaved the all-pervading ether."[4] Such deep faith placed by brilliant men in the empirically unproven ether resulted from the obvious successes of the "classical" (as Newtonian ideas had come to be regarded) system and the world view it produced. Surely the ether had to exist to make physics whole again.

▶   *Is the ether a universal reference frame?*

Maxwell had successfully predicted the speed of light as being about 186,000 miles per second.[5] But what was that speed to be measured against? Speeds are always *relative* to some measuring instruments – fence posts, racetrack lengths, or yard sticks – some "reference frame." In Newton's mechanics, the speed of a ball, for example, would normally be measured with respect to the person who threw it and the ground on which he stood.

But consider a man throwing a ball inside a large empty room in a ship which is moving with constant speed and a constant direction parallel to a shoreline. If the man throws a ball across the room, towards the front of the ship, the speed of that ball could be measured from two reference frames – with respect to the ship, or with respect to the shoreline. An observer on the shore could calculate the speed of the ball with respect to the reference frame of the shore. His figure for the speed would be higher than that calculated by the man on the ship using the reference frame of the ship. The difference would be just the speed of the ship itself.

The "relativity" of speed, that is, its dependence on the reference frame of the measurement, was an accepted fact and caused no problems for physicists as long as they were consistent in their measurements. Newton's laws had been found to work equally well regardless of which reference frame was chosen, as long as that reference frame did not change either its speed or direction during an experiment. In physics, the term "velocity" is used to signify both speed and direction, so the requirement for a reliable reference frame is that it have a "constant velocity." [The words "speed" and "velocity" are used interchangeably in common speech, and we will call attention to a distinction between them only when it is critical.] Such a smoothly moving frame is also called an "inertial" reference frame.

All experiments performed inside a ship that is moving with a constant velocity (an inertial reference frame), would obey Newton's laws as applied by observers on that ship, even if their measurements of time, distance, and velocity took no cognizance of the ship's motion. Observers on land (another inertial reference frame) would add the velocity of the ship itself to every measurement, but could use those different measurements in exactly the same Newtonian equations, and would find their calculated predictions in agreement with what they observed. Measurements may be relative, but the laws of physics cannot be relative. They must be exactly the same for all inertial reference frames. This principle, formulated by Galileo, was known as Galilean relativity.

It might appear that Galilean relativity, which stressed the universality of the law and the relativity of the measurement, was a curious reversal of priority. One might think that the only *real* results, the "true" velocity of the ball, for example, are the measurements made from the shore, the non-moving reference frame. The ship is moving, and it might be said that observers on the ship are ignoring reality if they do not include the velocity of the ship in their calculations. But the principle of Galilean relativity was essential for physicists, not because experiments were often conducted on ships or trains, ignoring the motion of the reference frame, but because from the time of Copernicus scientists had recognized that *all* experiments they performed took place on a giant moving platform – the Earth itself.

At the equator, the Earth's revolution about its axis moves scientists, balls, and oceans alike at nearly 1,000 miles per hour (and at smaller speeds at higher latitudes). The Earth as a whole is whizzing through space at 18 miles per second along its yearly orbit about the sun. So which is the "true" velocity of that ball? The one with respect to the ship? the shore? the center of the earth? the distant stars?

The answer, according to Galilean relativity, is that because velocity is *relative*, there is one equally true value for *each* reference frame. No one frame is privileged. The universally "true" aspects of physics are in the forms of the laws, the equations, and not in the individual measurements of velocity. And indeed the same Newtonian laws worked for all observers, on different places on the Earth, and at all points on the Earth's orbit. Fortunately the Earth moves smoothly in space, and its changes in speed and direction are slow enough so that for terrestrial physics the planet serves nicely as an "inertial" reference frame. Thus thanks to Galilean relativity, the laws of physics learned on the Earth could be universally true, and not in error just because the velocity of the earth itself had not been taken into account.

Now we return to Maxwell's equations, which had predicted a single value for the speed of light, 186,000 miles per second. But because velocity is a *relative* property, the following question arose: with respect to which particular reference frame is that figure of 186,000 miles per second the correct value? Newtonians were certain that this speed, as all others, was only one possible value, correct only with respect to one particular reference frame.

Like the moving ball, whose velocity could be measured with respect to either ship or shore, wave velocities must also be measured with respect to one or another reference frame. Water waves move at a fixed velocity with respect to the body of water which carries them, but can be observed to move at a different velocity with respect to the shore if there is a current moving that whole body of water along the shore. Sound waves in air are measured moving at one velocity with respect to the air itself, but at different velocities with respect to the ground depending on any wind (movement of that body of air).

So what was the proper reference frame for Maxwell's value of the velocity of light waves? Was that the correct value with respect to a reference frame attached to the source of light? Attached to the air? The Earth? The stars? With the ether theory, an answer was at hand: just as water waves travel at a fixed speed with respect to the medium of water on which they move, light waves must travel at that fixed speed, 186,000 miles per second, with respect to the medium which carries electromagnetic waves, the ether. The ether had again helped join Newtonian mechanics with Maxwell's electromagnetism. And there was at least one universe-wide reference frame: the all-pervading ether.

▶ *The Michelson-Morley measurement*

Physicists expected that if they measured the speed of light taking as a frame of reference the surface of the Earth, they would get varying results, because the moving Earth had to be constantly changing its velocity with respect to the universe-filling ether. Imagine we are measuring the speed of a beam of light coming from a distant star. At the moment we make our measurement, the Earth occupies some position in our orbit about the sun, and is moving at 18 miles per second (the Earth's average orbital speed) in some direction. Suppose that at the moment of our measurement, the Earth's motion is carrying it directly towards the star which is the source of light. Then the velocity of that beam, as we would see it relative to our reference frame attached to the Earth, should be *greater* than 186,000 miles per second (m.p.s.) by just the 18 m.p.s. with which the Earth was heading into the beam.

But six months later, the Earth would be halfway around the sun and moving in the opposite direction, so we would be moving away from the source of the light, in the same direction as the beam. Now the velocity of the Earth would have to be subtracted from that of the light, and we should observe the speed of the light as *less* than 186,000 m.p.s. by 18 m.p.s. Thus the speed of the light should vary up and down every six months by this small amount, about a 0.01% change.

Making an accurate measurement of the speed of something so dazzlingly fast as light was a great challenge. Experimental physicists, such as the ingenious American Albert Michelson, devised increasingly accurate means for making just such measurements. By the final decade of the 19th century, Michelson, working in Cleveland, Ohio with colleague E. W. Morley, had invented an instrument so sensitive that they could easily detect the expected plus or minus 18 m.p.s. change in the relative speed of light.

Incredibly, Michelson and Morley kept getting the same single value for the speed of light, no matter where they set up their instruments, no matter which way the light was moving, and no matter what the time of year. It was as if the ether stuck to every instrument, or to the Earth itself. Somehow, we were fixed in the ether, while the sun, planets, and stars moved about us.

Two centuries after Isaac Newton had won Copernicus' battle to dislodge the Earth from dead center in the cosmos, physicists seemed to be finding the Earth back in a unique position, that of being the one body in the universe at rest in the all-important ether. But scientists would not accept a return to an Earth-centered universe. The result of the Michelson-Morley experiment, repeated with increasing precision but yielding the same result on through the 20th century, was instead to doom the attempt to reconcile Newtonian mechanics with Maxwell's electromagnetism.

▶   *Einstein's solution*

The young European Einstein may not have known of the American experiment by Michelson and Morley; neither his later memory nor the written record are clear on that point.[6] For whatever reason, Einstein concluded about 1903 that the ether was a hopelessly clumsy construct. Electromagnetic waves had to be waves in some medium only if the analogy with Newtonian sound or water waves was to be maintained. There was no such strict necessity, only a desire for a medium to blend Maxwell's electromagnetic waves into Newton's mechanical scheme.

Einstein, who had rebelled against authority as a student, felt free in his mid-20's to reconsider the entire body of physics. To create his own synthesis, he selected those parts he found satisfying philosophically. Experimental evidence was always crucial, but Einstein bound himself, throughout his life, only to the letter of experimental law. As long as the directly observed data were incorporated, Einstein felt free to establish theories on any plan, whether in keeping with traditional approaches or not. In this case he found Maxwell's equations more beautiful and less restrictive than Newton's. With his astonishing 1905 paper, "On the Electrodynamics of Moving Bodies,"[7] Einstein reduced Newtonian physics to a convenience, highly valuable for small-scale physics, but far from the spirit of the laws through which the universe operated.

Einstein's paper, which announced what became known as the Special Theory of Relativity, begins with two assumptions, two convictions about how the universe works. These assumptions, from which all the astonishing predictions of the theory result, are themselves straightforward. The first is that when we find the correct laws of physics, those laws will obey the central tenet of Galilean relativity: they will have exactly the same forms in all uniformly moving, "inertial," reference frames.

The second postulate is even simpler. Maxwell's equations had predicted a speed of light, but without specifying a reference frame. Einstein accepted that lack of specification as meaning that no specification is needed. This second postulate is that the speed of light will be observed to be the same for *any* inertial frame. No ether frame is necessary. Measuring the speed of light offers no clue as to whether or not an ether exists, because any inertial frame can be used with the same resultant speed. Thus the Michelson-Morley experiment detected only the one, universal value for the speed of light. As long as the Earth moved uniformly, even much larger changes in motion between measurements would have made no difference whatsoever. Slower velocities (sound, water waves, balls) remained relative, but the speed of light became a new absolute quality of the universe.

▶ *Why Einstein's postulates are strange*

These two postulates do not seem at first the kind of radical departure that would create a revolution in world view. The inescapable implications of the postulates, however, are that while Maxwell's equations of electricity and magnetism are satisfactory descriptions of the world, Newton's equations of mechanics are not. Equations of mechanics based on Einstein's two postulates are entirely different from Newton's equations, although the Einsteinian versions come exceedingly close to Newton's for objects whose relative speeds are small compared to the speed of light. Causing even more fundamental upheaval was the realization that Einstein's postulates necessitate entirely new conceptions of time and space.

Imagine a beam of light from a distant star skimming over the surface of the Earth. Observers on the Earth measure the speed of the beam at 186,000 miles per second. A spaceship is seen chasing after that beam of light, i.e. moving in the same direction as the beam. Suppose the spaceship's velocity, measured from the Earth, is 185,000 miles per second. What velocity will the beam of light have, if that velocity is measured relative to the frame of reference of the spaceship? Only 186,000 - 185,000, or 1,000 miles per second, would be the prediction made by observers who assumed Newton's rules for relative velocities. But according to Einstein's second postulate, the spaceship would have to measure 186,000 miles per second, not 1,000, for that beam of light it's chasing. How can the same beam of light have the same velocity as seen from these differently moving references? If the spaceship did indeed get the same answer as the Earth-based observers, as Einstein predicted, then common sense would dictate that one measurement must have been an illusion of some kind, a systematic flaw in the measuring apparatus.

Just before Einstein's theory was published, a partial explanation involving just such "flawed" measuring instruments was developed by Lorentz and FitzGerald.[8] Their work supposed that instruments moving through the ether would be compressed, or shrunk along their direction of motion in the ether. This shrinking would lead to erroneous measurements, by just the correct amount so that these moving observers would measure light speed as if it remained constant, despite their own motion. Observers on the spaceship in the previous paragraph would measure the light beam as travelling a full 186,000 miles in one second relative to the ship. The beam actually would travel only 1,000 miles in that second, relative to the ship. The erroneous measurement would be due to the contracted instruments on the ship. Their rulers (of whatever design) would actually be only 1/186th of their correct length, making the 1,000 miles appear to be 186,000 miles to their instruments.

The "Lorentz-FitzGerald contraction" would solve the immediate difficulty of the Michelson-Morley experiment, because if the Earth itself moved through the ether, all earth-bound instruments would be contracted just enough to mask the effect of that motion. Only observers at rest in the ether would have the true lengths, and could make accurate measurements unhampered by shrunken measuring devices. Observers on moving frames might not realize anything was wrong, since all their measures would be off in one direction by the same amount, but when they measured lengths of objects moving more slowly or even fixed in the true rest frame, those objects would appear *stretched* to the contracted instruments of the moving observers. Thus some objects with some motions would appear longer, while others with other motions would appear shorter. The true ether-frame rest system would be the one that saw *all* motion causing only contraction.

Something like the Lorentz-FitzGerald contraction can be derived from Einstein's theory, but Einstein's version is both more subtle and surprising. Simple contraction for observers moving with respect to the true ether frame would violate the strict letter of Einstein's first postulate, for *all* inertial observers must have equal access to truth using their instruments and the same laws of physics. In the Lorentz-FitzGerald version, only the ether frame observers get true measurements. In Einstein's version, *any* observer will see *any* object moving relative to himself to have shrunk in the direction of apparent motion. The entire space of moving frames, not just the objects in them, will be observed as contracted in this one dimension. But no frame will be able to make any special claim to privilege. *Every* frame will see *every other* frame as contracted in the direction of motion.

This contraction is not like any phenomenon of common experience. First, the contraction is not a mere ordinary mechanically forced distortion of physical objects, but a change in space itself. Not only measuring instruments, rigid bodies, atoms and planets, but distances between these atoms and planets would all be observed as contracted in the direction of relative motion by any observer watching these things move by. Consider watching a 12-inch ruler pass by at 86 per cent of the speed of light, moving exactly along its length. The observer would note that the height and thickness of the ruler remained identical to those dimensions at rest, but now the length would have been reduced. The 12-inch ruler would be uniformly contracted to only 6 inches as measured by the observer. All twelve inch marks would remain, but with only half the distance between them as before. The numbers would be skinny, half their former width, and the marks themselves would be thinner by half. The ruler would be under no mechanical compression or strain, however, and all objects and processes moving with it would behave properly. A watch, with formerly round gears now appearing oval shaped and contracted only in the direction of motion, would still tick smoothly. The second hand would appear to contract and re-expand as it rotated into directions along and then across the direction of motion.

A second conventional explanation for such strange observations would be that a visual distortion, an optical illusion, has taken place. Relativistic space contraction, however, is independent of the specific means of measurement, and thus cannot be merely a trick of the eye or the camera. Every means of measurement, using optical, mechanical, sonic, or other sensing devices, would record exactly the same results, that objects and distances in any frame moving relative to the measuring apparatus are contracted, shrunk, in the direction of motion. There is no avoiding the conclusion that this space contraction is an actual phenomenon to the full extent that reality is accessible to measurement.

In practice, there is at present no way to achieve sufficiently high relative speeds so that the contraction of people or nearby objects would be significant enough to be detected visually. Even if such speeds were reached, direct visual observations of the Lorentz contraction would be greatly complicated by more conventional distortions in appearance caused by the finite travel time of light. If an object passes the observer across his line of sight, light from one end of that object would reach the observer's eye sooner than light from the other end, depending on the differing distances of each end from the observer. This would result in a complex visual effect added to the actual length contraction.[9]

As another example of what measurements would reveal if extremely high speeds were achievable for nearby large objects, imagine we are observers on Spaceship A watching twin Spaceship B speed past at nearly the velocity of light. We would observe, confirmed by all our measuring instruments, that Spaceship B was contracted from its original length. Observers on Spaceship B, nevertheless, would report that their spaceship was perfectly normal. Spaceship B's crew would look out and see Spaceship A contracted, as confirmed by all the instruments on Spaceship B. The symmetry is perfect; each ship sees the other's space contracted by exactly the same amount. The faster the relative velocity, the greater the contraction, but only the *relative* velocity matters. There is no absolute reference frame needed for comparison, and the speeds of the ships with respect to the Earth, or the galaxy, have no effect on their mutual observations of each other. This symmetry was completely lacking in the original Lorentz-FitzGerald hypothesis.

But this mutual observation of contraction seems logically contradictory. We saw their spaceship contracted; they saw ours contracted. Who really shrank? Both? The answer is that each one *really did shrink* with respect to the other's frame of reference. And, each one *really did remain unchanged* in its own frame of reference. Length has reality only with respect to some frame of reference. By every conceivable observation, we on Spaceship A find that Spaceship B has shrunk, and ours has not. By every conceivable observation, Spaceship B's crew concludes that they did not contract, but that Spaceship A did. Both are completely correct. Einstein's new physics does not provide for any higher claims to the reality of length than the results of every conceivable measurement.

Even if these results are internally consistent, they conflict with common sense. A deeper understanding of time and space emerges from this apparent contradiction of observations. Our sense of offended propriety is based on our own experience on the Earth, where two observers' relative motion hardly ever exceeds the speed of sound, and never approaches the speed of light, about a million times faster than sound.

Because of the complete reciprocity of length contraction between any one inertial frame and any other, Einstein concluded that just as velocity had always been regarded as relative, now space must also be regarded as relative to the observer.

But space contraction was not enough. Time had to become relative as well, to avoid other contradictions in measurement. A formula for time slowing, or time "dilation," similar to that for space contraction, came out of Einstein's postulates. Clocks, heartbeats, biological growth, and chemical reactions in one frame are all slowed down as observed from any other frame moving relative to the first. Astronauts in a spaceship moving near the speed of light, relative to Earth, age more slowly, as seen from the Earth. The faster the speed, the greater the time dilation. As a moving object approaches the speed of light, its time is observed to slow to an infinitesimal crawl. This phenomenon too is perfectly symmetric. A spaceship would find its own time steady and normal, and would observe the Earth whizzing by turning more slowly, and Earth clocks and people slowed down. The spaceship is correct; and so is the Earth. In each case the other's time *has* slowed down. [A variation on this situation is a famous puzzle known as the "twin paradox," discussed in Chapter 4.]

Mass soon joined the list of relative properties along with space and time. An object of given mass in one frame will appear more massive if viewed from another frame which is moving with respect to the original. The faster the motion of a given object, the greater its gain in mass, until the mass appears to approach infinity as the speed of light is neared. Indeed, all three variable quantities – space contraction, time dilation, and mass increase – become infinite for frames moving at the speed of light. A material object would shrink to infinitesimal thickness, its time would slow to a stop, and it would increase to infinite mass. These extremes are unattainable in our universe. The property of increasing mass, for example, means that to speed a body up becomes increasingly difficult as its velocity nears that of light. All finite energy sources would be exhausted as the mass of the body increased without limit, so the speed of light itself could never be reached. Only light can be observed at that speed.

▶   *The myth of the universal present*
Simultaneity is a profound property which also becomes relative in Einstein's theory. In many ways this change is more startling than relative time or space themselves. Two events with any finite spatial separation, such as snapping the fingers of both your hands with arms outstretched, have two individual sets of space and time values for any observer. Common sense and classical physics have no difficulty in classifying those two events universally as either simultaneous or not simultaneous.

In special relativity, however, simultaneity is relative to the reference frame. Fingers that snap simultaneously for observers on the Earth do not snap simultaneously as seen from a spaceship moving parallel to your outstretched arms. From one moving frame, your left hand fingers might have snapped before the right. From a frame moving opposite to the first, the right snapped before the left. Although Einstein's laws protect causality, so that the order of two events causally linked is always observed to be the same, independent events separated in space will appear to happen in reversed order for oppositely moving observers.

The existence of absolute, universal simultaneity is a deep cultural value. Many creation myths and social world views assume that the universe has some particular existence for every moment of time, even if humans are unable to perceive more than a miniscule fraction of that universe. The failure of simultaneity to be an absolute property implies that "the universe at one moment" has no verifiable reality. Moments are not universal; the present is a parochial concept, valid for each observer, but with a different meaning for any observer in any other inertial frame.

For situations such as the snapping of fingers, the failure of simultaneity is of such small magnitude that the difference would never be noticed, even when comparing views from spaceships moving near light-speed. In astronomical situations, however, the failure of simultaneity changes drastically the meaning of any description we try to make of events on a galactic scale. A cluster of galaxies, such as shown in Fig. 2, covers a gigantic volume of space. Every extended smudge in this picture is an entire galaxy of billions of stars, each comparable to our sun. Our galaxy, the Milky Way, if viewed from somewhere in this cluster, would also look like a tiny smudge. Our sun would be one point of light, lost amidst the 200 billion other single suns in our smudge of light.

The spatial scale of this picture is immense. From a smudge of light representing a galaxy on the left to one on the right is a measure of millions of millions of trillions of miles. In terms of the distance light travels in one year (six trillion miles), called one "light-year," the galaxies in these pictures are millions of light-years apart. The cluster is also immensely distant from us, so the light captured in this photograph took an average of about 400 million years to reach the camera. Light from the more distant galaxies in this picture might have taken an additional time, say 10 million more years. This immensity of time-span is awesome, but that time-span in itself is not the whole story.

Because of these huge, various distances, this picture is not a snapshot of a cluster of galaxies at one moment, 400 million years ago, in the history of the universe. This photograph is actually like a collage: a montage of photographs of the various galaxies in the cluster at different moments in time, moments millions of years apart. That fact was understood in Newtonian physics.

In classical physics we could, at least in principle, assemble a true photograph of the cluster's members at one particular moment. From this photograph we would cut out and save the image of the

Fig. 2. Cluster of galaxies in the direction of the constellation of Hercules. © 1973 National Optical Astronomy Laboratories/Kitt Peak, reproduced by permission.

furthest galaxy from us. That image shows the galaxy as it looked about 410 million years ago. The next galaxy is only about 409 million light-years away, so its image on this photo is only 409 million years old. So we would have to wait an additional one million years, to photograph the image of that second-most-distant galaxy to match the age of the image of the most distant member of the cluster. By repeating this process, re-photographing each galaxy at the proper time, we could eventually assemble a new montage, showing the appearance of the entire cluster at one moment in time, 410 million years before our first photograph.

The classical physicist would have an additional minor difficulty to resolve in order to accomplish this task, beyond the long duration of the project. These galaxies are moving, along with our own, so that a correction for the changing distances would have to be calculated to make the photographs at the proper time. But again, at least in principle, this procedure would have worked.

According to Einstein, however, the idea of a universally "correct" photograph of a cluster at a moment in time is meaningless. Our photograph of a moment, 410 million years ago, that was so painstakingly assembled to correct for the finite travel time of light, would still be "correct" only for that tiny part of the universe moving in our own space-time reference frame. For any other inertial frames, including those of other stars in our galaxy, or of the galaxies in the cluster we are photographing, our new composite photograph would be just as inaccurate as the original photo. Because simultaneity is not universal. A "moment" for other observers is a *different* montage than ours, a different mix of those millions of years. And their montage has an equal claim to reality.

The idea of a universal present is so important that it should be afforded the status of a myth. Deeply ingrained in western world views had been the concept of time flowing uniformly all over the universe. The state of that universe could be known, at least to God, as it changed from instant to instant. But now the "universe as a whole" has been separated into fragments that can never share a universal moment of time. Even the "moment of creation" could not exist, unless the creation of the universe happened at a single point in space.[10]

Science had relied on that component of reality, universal time, and it was a central feature of Newton's physics. In Newton's words,

> Absolute, True, and Mathematical Time, of itself, and from its own nature flows equably without regard to any thing external.

Space followed suit:

> Absolute Space, in its own nature, without regard to any

thing external, remains always similar and immovable.[11]

Within the assumption of absolute space and time came a corollary: that space and time are independent of each other and of the objects in the universe. Einstein noted that Newton "quite explicitly" introduced absolute space "as the omnipresent active participant in all mechanical events; by 'absolute' he obviously means uninfluenced by masses and their motion."[12] There had been no reason to doubt the independence and absolute nature of time and space.

The dilation of time, contraction of space, increase of mass, and failure of simultaneity, all had to be confirmed by difficult experiments, since these phenomena required large speeds before they became large enough to observe. The development of sub-atomic physics in the 20th century has given physicists the ability to observe tiny particles travelling near the speed of light. By the centennial of Einstein's birth, the predictions of Special Relativity had been confirmed in thousands of experiments. Even before these experimental validations, however, the loss of absolute space and absolute time had opened new theoretical possibilities. Space, time, mass, electricity, magnetism, gravity, and other properties, formerly thought to be independent by virtue of absolute status, might actually interact. For example, mass could conceivably affect time and space by its presence. That theoretical potential was realized only a decade after the Special Theory, with Einstein's masterwork, the General Theory of Relativity.

▶   *The general theory*

In his much more complex and difficult General Theory of Relativity (1915),[13] Einstein extended considerably the principles he had put forth ten years earlier. First, he said the laws of physics must apply to *all* systems of reference, whether inertial (at rest or moving uniformly) or moving with changing velocity. The Special Theory dealt only with inertial systems. In the General Theory Einstein was able to amalgamate inertia and gravitation into a single concept, thus dropping the classical distinction between a body's inertial motion and its motion under the action of gravitation.

The term "mass" had two quite separate meanings in classical physics. "Inertial mass" was that property by which a body resisted any change in its state of motion. It is hard to start a large mass moving, and once it is moving, just as hard to slow it down. The resistance to change is proportional to the amount of "inertial mass" the body has. The other mass was "gravitational mass," a measure of the strength with which the force of gravity acts on a given body. The larger an object's "gravitational mass," the more strongly it is attracted by the Earth's gravity. These are independent definitions of the word mass, and there is no inherent reason to suppose that there is any connection between these two properties. Why couldn't a body of large inertial mass have small gravitational mass?

But as Galileo's famous experiment of dropping a heavy and a light ball off the tower of Pisa[14] was designed to demonstrate, the two types of masses were found *always* to be equivalent. One origin of Einstein's General Theory of Relativity was a distaste for the apparently coincidental nature of this equivalence. Einstein succeeded in making the identity of inertial and gravitational properties an inherent feature of his scheme for the universe. In the process, gravity, space, and inertia lost their independence and became a part of a grander concept of warped space-time.

In the General Theory, inertial reference frames *in gravitational fields* (for example in an area near a mass like the Sun) and *accelerating reference frames* (those with a changing velocity) are equivalent. In Einstein's popular example, a person in a closed elevator cannot in principle tell whether the pressure on his feet is due to a gravitational field or to an upward acceleration of the elevator. If he tosses a ball across the elevator, it will follow a curved path which would be the same whether caused by acceleration of the elevator or by the action of a gravitational field of a certain strength. Einstein describes this equivalence in his "Autobiographical Notes." "In a gravitational field things behave as they do in a space free of gravitation, if one introduces in it, in place of an 'inertial system,' a reference system which is accelerated relative to an inertial system."[15]

New mathematics turned out to be helpful in describing the new interrelated space and time, and a new language became part of culture: the three dimensions of space, and one of time, merged into four numbers of the new space-time. In four-dimensional geometry, motions describing inertial systems are represented by straight lines (called world lines), but in accelerated systems, these world lines become curved. "The concepts 'straight' and 'curved' become relativized, in so far as they refer to orbits of the light rays and of freely moving bodies. Through this, the whole structure of Euclidean geometry is made to totter. For it rests essentially on the classical law of inertia, which determines straight lines."[16] Einstein asserted that the geometry of the real space-time continuum is non-Euclidean or curved. The presence of mass distorts space itself, so that a straight line near a large mass is no longer the shortest path between two points. By way of rough analogy, the shortest distance between London and New York on the curved surface of the Earth is a curved line. A straight line leaving London would go off into space or down into the earth, and would have to be connected to another straight line in order to reach New York. A curved line is thus the shortest distance between two points on a curved surface.

The flavor of this new formulation can be appreciated from the new way in which gravity was described. For Newton, gravity was a force between any two masses. For Einstein, any mass distorts the space-time around itself. If another mass is present, it too distorts space-time, and then the two masses move in their mutually changed space-time environment. The warped field of space-time determines the motion of the masses, replacing the action-at-a-distance force of gravity described by Newton. Newton's law of acceleration due to force is replaced by Einstein's field equations, describing the mutual interactions of masses and space-time.

From the General Theory one can describe planetary orbits as "geodesic" lines. A geodesic line in geometry is the shortest possible line between two points. On a two-dimensional flat surface, that geodesic will be a straight line. On the two-dimensional surface of a three-dimensional globe, the geodesic will be a great circle. The curvature of planetary orbits, which classical physicists regarded as due to gravitational force, becomes in the new theory a result of the curvature in four-dimensional space-time itself. A planet, in other words, travels the path it has to – the only one it can follow – because of the curved space caused mainly by the sun's gravitational field. This path was slightly different from that predicted by Newtonian physics, and at last physicists and astronomers had a physical theory which explained precisely the deviation in Mercury's orbit from Newton's predictions.[17] Of all the planets, only Mercury, closest to the sun and hence in the region of greatest curvature of space-time (the strongest gravity in classical language), deviated sufficiently from a Newtonian orbit to require the use of Einstein's physics to explain its motion.

Since world lines become curved or geodesic in an accelerated or gravitational system of reference, Einstein asserted that even light must travel in a curved path. The English astronomer Eddington made the first test of this assertion in 1919 when he photographed stars near the sun during a total eclipse. These stars seemed to be in slightly different positions from those they held when the sun was in a different part of the sky. This displacement of the apparent positions of the stars demonstrated that their light had been "bent" as it passed through the sun's gravitational field, and the bending was just the amount predicted by Einstein.[18] Light follows the curvature of space. The geometry of space-time changes when matter is present. As Louis de Broglie explains, "Space-time is not Euclidean in the presence of a gravitational field; hence, as for a curved surface, it is impossible to adopt Cartesian coordinates. Gravitation is thus reduced to an effect of the curvature of space-time dependent upon the existence of masses scattered through the universe."[19] Since matter is present in various places throughout the universe, gravitational fields are also present which curve the light rays passing through them.

Einstein's propositions about the structure of the universe gave cosmology a new scientific underpinning. Through mathematical deductions from his General Theory, he said that one possibility was that space-time was non-Euclidean, and finite, but *without* boundaries. Classically, a traveler on a sphere might keep going indefinitely in any one direction, but would always return to his starting point. He would find himself exploring a surface of definite finite size, but never meeting a boundary which blocks his path or closes off the surface. By analogy, light rays in a finite but unbounded universe travel geodesic lines and eventually return from the opposite direction to their source.

Following from the General Theory of Relativity are today's concepts of an expanding universe which originated in a "Big Bang," and of such astonishing "inhabitants" of this universe as pulsars, neutron stars, black holes, and gravity waves. Such a cosmology contrasts sharply with classical concepts of infinite space describable in terms of Euclidean geometry. As Einstein pointed out, "We cannot use in the general relativity theory the mechanical scaffolding of parallel and perpendicular rods and synchronized clocks.... Our world is not Euclidean. The geometrical nature of our world is shaped by masses and their velocities."[20]

Newton's laws had been accepted for so long because they were successful in so many realms. Thus Einstein's physics was obliged to resemble Newton's in those realms of slow speeds and weak gravity where Newton's laws worked well. In physicist Eric Rogers' phrase, "Where the new cloth meets the old cloth, they *must* agree."[21] This *correspondence* between old and new has become a guiding principle in all 20th-century physics.

The Newtonian law of gravitation is now regarded as a special limiting case of the General Theory of Relativity. In other words, the General Theory includes the old gravitational law and goes well beyond it. Newton's laws are still eminently useful in a vast set of situations, ranging from calculating the path of a baseball to planning a flight to the moon. In all of these cases, Einstein's physics is essentially identical to Newton's. But if the gravitational forces are strong, as near the sun's gravitational field, the old law gives wrong answers. Newton's old theory of universal gravitation must be superseded by Einstein's new one.

Einstein's General Theory of Relativity presents almost insurmountable mathematical difficulties for lay readers. Few outside the physical-mathematical field can follow the mathematical language of Maxwell's equations, let alone some of the mathematics involved with General Relativity. Feynman points out that in addition "as we go to more and more advanced physics, many simple things can be deduced mathematically more rapidly than they can be really understood in a fundamental or simple sense....There are circumstances in which mathematics will produce results which no one has really been able to understand in any direct fashion."[22] The mathematics of "gravitationally collapsed objects" had been known for several decades before an understanding of how to look for them astronomically developed, and before the evocative name "black hole" was applied. Physicists often are quick to note that mathematicians do not have to deal with reality. Their constructs and systems do not have to correspond to experimental results. Gibbs quipped that a mathematician may say anything he pleases, but a physicist must be partially sane.[23] Mathematics, nevertheless, has been an indispensable tool for physicists. Newton had to discover the fundamental theorem of calculus (which he called "fluxions") before he could exploit his laws. Partial differential equations were developed before the physicists learned to apply them to electromagnetic fields. And Einstein depended on Gauss, Riemann, and other mathematicians for the geometry of his space-time system. What has happened to 20th-century physics is that much of the theory can be described only in abstract mathematical terminology. The old pictorial and mechanical models cannot adequately describe what is happening in reality.

▶ *Lessons from the new physics*

The practical applications of the Special and General Theories of Relativity are still limited to the work of physicists and astronomers. The wondrous and strange phenomena of space contraction, time dilation, and mass increase have had no effect on the needs of mankind for food, clothing, or shelter. This isolation contradicts the widely perceived and influential notion of a crucial link between Einstein's physics and the practical release of atomic energy: that faulty connection will be discussed in Chapter 6.

The attempt to find moral or philosophical applications of Einstein's relativity has been widespread. What lessons might a culture have derived from the new physics? That one must be wary of "common sense" based on too limited an experience in the cosmos; that two observers might describe the events of the world differently, and yet (barring blunders) each might be correct; that the universe obeys rigid laws, as the Newtonian clockwork universe did, but that those laws are far more subtle and interwoven than had been imagined.

Other notions have also been derived from Einstein's physics, notions that nearly all philosophers of science and physicists, including Einstein, would agree have little or no relation to the actual physics. Yet these have become far more pervasive than the list above. These notions are that every measurement is subjective; that all "truth" is only relative; that science has abandoned cause and effect; that everything is relative.

The relativity of truth is an erroneous conclusion that arises all too easily out of the accurate statement that measurements of time, space, and mass are relative to the individual observer's space-time reference frame. But in fact nothing is subjective or uncertain about those measurements. The numbers must agree for all observers sharing a reference frame. Any measurement can have one and only one true value for each reference frame. Observers in a second reference frame must not only agree with each other, but, using the new laws, their single, correct measurement can be predicted completely from the first frame's measurement. Thus the relativity of these measurements is no more an abandonment of absolute truth or an adoption of subjectivity than the older relativity of velocity alone had been. Furthermore, the speed of light, and the new laws of physics, are universal, absolute truths in the Theory of Relativity. The net effect, then, should be to declare some previously absolute characteristics to be objectively and predictably relative, while establishing new absolute laws and measurements.

Finally, cause and effect are rigidly maintained in Einstein's physics. A great world-machine like Newton's clockwork is still possible, but mechanical models for the clock are no longer adequate. A mathematical machine is necessary, so if the image of God as a clockmaker is not appropriate, a new image of God as a computer programmer could be fashioned instead.

Despite the continuing protests of the physicists and philosophers, "everything is relative" is still a phrase heard in proximity to Einstein's name. In the chapters to come, we shall examine the vast range of cultural phenomena for which Einstein serves as muse, including many which interpret his work with subtlety and grace, as well as those in which the diametrically opposite "everything is relative" provides a license for claims of a relativism of truth and morality that Einstein himself despised. Except for its name, the Theory of Relativity remains a monument to a faith in the ultimate certainty of knowledge.

4

Einstein becomes a muse

In the beauty of poems are the tuft and final applause of science.
Walt Whitman

The Newtonian clockwork is smashed; time and space are relative to the observer; universal moments of time are mythical: these are the messages which burst into Western culture with the headlines of November 1919. With the positive results of the eclipse expedition came the first public awareness of Albert Einstein and any of his work. An intense demand for popularizations marked the general hunger for news of this revolutionary view of how the world worked. Both accurate and misleading versions of Einstein's work were diffused, and broader cultural effects appeared within months of Einstein's instant fame.

Writers paid attention when such journals as *The Dial, Current Opinion, Harper's, Contemporary Review, The Living Age*, and *The New Republic* took considerable effort to explain the new physics. In addition to the general excitement in the press, outlined in Chapter 1, many articles contained provocative challenges to apply relativity beyond physics itself. In an article headed "The Most Sensational Event in Physics Since Newton: Father Time Goes Out of the Picture," *Current Opinion* (January 1920) said of the theory, "Fantastic consequences flow from it, consequences quite inconsistent with the traditional philosophy of science."[1] In *Harper's*, an illustrated article headed "A New Conception of the Universe" challenged readers: "Once more we must seek to overcome mental inertia, to liberate ourselves from preconceived ideas."[2] *The New Republic*, after two long and detailed expository essays,[3] proclaimed a new era:

> Despite the insistence of intellectual mediocrity that the proper study of mankind is man, nothing is of such truly human interest as the nature of the physical world in which we live. Moreover, to gain a genuinely new fundamental idea

such as is involved in the theory of relativity is an experience akin to that which comes in the highest creative art or religion – a liberation from the dead complacency of the accepted views and an enlargement of our being by an enlarged vision of new possibilities.[4]

► *Positive poetic responses to the muse*

▷   William Carlos Williams
The notion of Einstein and his work as literature is echoed in William Carlos Williams' "St. Francis Einstein of the Daffodils" (reproduced in full in an appendix to this book). In perhaps the first poetic manifestation of Einstein as muse, Williams presents the physicist as a savior bringing freedom from the "dead" and "old-fashioned knowledge."

> April Einstein
> through the blossomy waters
> rebellious, laughing
> under liberty's dead arm
> has come among the daffodils
> shouting
> that flowers and men
> were created
> relatively equal.
> Oldfashioned knowledge is
> dead under the blossoming peachtrees.[5]

This poem appeared in *Contact IV* (1921) following Einstein's visit to America in April 1921. In its revised version in *Adam & Eve & the City*, it is subtitled "On the first visit of Professor Einstein to the United States in the spring of 1921."[6]

Certainly Williams was well aware of the popular press coverage of Einstein and relativity, as "St. Francis Einstein of the Daffodils" indicates:

> Sing of Einstein's
> Yiddishe peachtrees, sing of
> sleep among the cherryblossoms.
> Sing of wise newspapers
> that quote the great mathematician:
> A little touch of
> Einstein in the night –[7]

Even if Williams did not read the extensive articles about relativity in *Science* or *Scientific American*, he probably read the *New York Times* and some of the popular journals mentioned above. Certainly he saw an article by Thomas Jewell Craven entitled "Art and Relativity" in *Dial*.[8]

How much of relativity did Williams actually understand and use in this 1921 poem? First, Williams seems to have appreciated Einstein's position that not all data true for one observer will also be true for an observer in another frame of reference. In the poem Einstein is both "tall as a violet/ in the latticearbor corner" and "as tall as a blossomy/ peartree!" depending on one's frame of reference. Both observations can be accurate, without contradiction. Second, as space and time become fused in relativity, the classical concepts of absolute three-dimensional space and separate absolute time must be abandoned. Williams writes that "at last, in the end of time,/ Einstein has come by force of/ complicated mathematics" – that is, in the end of the formerly accepted, absolute time. Multiple descriptions of events, which in a given three-dimensional world could not all be valid, coexist in a metaphoric four-dimensional relativistic world. The peartree sways "with contrary motions," and the "spring winds [are] blowing/ four ways, hot and cold."

Thematically the poem rejoices at the liberation Einstein brings – Williams associates the spring imagery of new life with an intellectual revolution: "Einstein has come/ bringing April in his head," bringing "spring-time of the mind." The natural world is also liberated; the physicist has brought "freedom/ for the daffodils/ till the unchained orchards/ shake their tufted flowers" as the physical world participates in and reflects the revolution in ideas. The orchard owner, "with windows wide," throws "off his covers/ one by one," in response to the promising change, much as the revolutionaries in all fields discarded the old conventions and opened up to new possibilities. The poem certainly welcomes the physicist and his theories with an enthusiasm and optimism quite opposed to the tone of *The Waste Land* (which Eliot wrote within months of the publication of "St. Francis Einstein of the Daffodils").

At this stage in his understanding of relativity, Williams had not realized the details of the relativity of measurements – a realization that was to lead him later into formal changes and a theoretical justification of his new poetic structure of the "variable foot." "St. Francis Einstein of the Daffodils" indicates that in 1921 he was interested in the new physics but that he had not created any formal poetic analogs to it. That came later with Williams' longer poem, *Paterson*.

Williams' essays and letters indicate that by the mid-1920's he had read Alfred North Whitehead's complex and precise *Science and the Modern World* and Charles Steinmetz's *Four Lectures on Relativity and Space*. He discussed the new physics with his engineer friend John Riordan and with the poet Louis Zukofsky, who translated a biography of Einstein. This enlarged knowledge of relativity provided Williams both subject matter and justification for his experiments with the variable foot.

Williams asked, "How can we accept Einstein's theory of relativity, affecting our very conception of the heavens about us of which the poet writes so much, without incorporating its essential fact – the relativity of measurements – into our own category of activity: the poem? Do we think we stand outside the universe? . . . Relativity applies to everything."[9] Insofar as a poetic line is a measure or group of measures, Williams considered it necessarily relativized. His new measure, or variable foot, developed partly from his understanding of relativity. "Poems," he said, "cannot any longer be made following a Euclidean measure, 'beautiful' as this may make them. The very grounds for our beliefs have altered . . . . Relativity gives us the cue . . . . We have today to do with the poetic, as always, but a *relatively* stable foot, not a rigid one."[10] [italics are Williams'] This connection Williams made between measurements in relativity and measures in poetry influenced many younger poets from Allen Ginsberg to Charles Olson.

In *Paterson*, Book Two, Williams discusses the problems of inventing forms, particularly poetic lines, which could give shape to the changing new content:

>Without invention nothing is well spaced,
>unless the mind change, unless
>the stars are new measured, according
>to their relative positions, the
>line will not change, the necessity
>will not matriculate: unless there is
>a new mind there cannot be a new
>line, the old will go on
>repeating itself with recurring
>deadliness . . .[11]

For Williams, the eclipse expedition's new measurement of the stars according to their relative positions generated a new mind, an appreciation of relativity which applied to the poetic line. The variable foot, as it appears in *Paterson*, measures a tripartite line in which the components vary according to contextual relationships: sentence fragments, single words, long phrases – all moving in a changing but "relatively stable foot."

Williams' conscious application of relativity to his poetry also helped him move beyond the stasis of Imagism to the pluralistic process of *Paterson*. His major formal problem in tackling a work the size of *Paterson* was to go past the preferred viewpoint and absolute time, which Imagism shared with classical physics, to the multiple viewpoints and relativistic space-times which could give form to the changing processes in the complex field of *Paterson*. In Williams' work, then, we find our first example of Einstein, through his personal fame and through popularizations of his work, serving as a muse supporting the continuing transformation of 20th-century poetry.

▷   Archibald MacLeish
        Interest in Einstein stimulated many writers and artists in Paris. Robert McAlmon recalls that especially during the mid-1920's "there was a period that was very Einstein or relativist . . . ." He wondered if time would "prove Einstein to have made such valuable discoveries as everybody so aggressively claimed in those days."[12] One member of the Paris avant-garde at the time was Archibald MacLeish, who published his long poem "Einstein" in 1926.[13] Hyatt Howe Waggoner considers this poem to be "not only the finest poetic tribute to the scientist, [but] also an informed and interesting comment on the philosophical significance of Einstein's achievement. The chief idea to emerge from the poem is that the new Einsteinian science has at once emphasized the centrality and increased the loneliness of the knowing mind."[14]

The poem opens with Einstein contemplating "finity": his own physical limits in space and time.

> . . .He is small and tight
> And solidly contracted into space
> Opaque and perpendicular which blots
> Earth with its shadow. And he terminates

> In shoes which bearing up against the sphere
> Attract his concentration . . .

But Einstein's mind does not share such limits.

> Nor could Jehovah and the million stars
> Staring within their solitudes of light,
> Nor all night's constellations be contained
> Between his boundaries . . .

After rejecting a hypothesis of common sense, simple three-dimensional reality,

> A world in reason which is in himself
> And has his own dimensions –

Einstein seeks unsuccessfully for images and models capable of expressing his ideas. But conventional models cannot portray a four-dimensional space-time continuum. He then turns to violins that "can sing/ Strange nameless words that image to the ear/ What has no waiting image in the brain." But the music "vaguely ravels into sound" and cannot be the language of his thoughts. Words do not work either. "There is no clear speech that can resolve/ Their texture to clear thought . . . . Now there are no words/ Nor names to name them." So the physicist moves into further abstractions until in his mind he comprehends the cosmos.

> He lies upon his bed
> Exerting on Arcturus and the moon
> Forces proportional inversely to
> The squares of their remoteness and conceives
> The Universe.

> Atomic.

> He can count
> Oceans in atoms and weigh out the air
> In multiples of one and subdivide
> Light to its numbers.

> If they will not speak
> Let them be silent in their particles.
> Let them be dead and he will lie among
> Their dust and cipher them – undo the signs
> Of their unreal identities and free
> The pure and single factor of all sums –
> Solve them to unity.

Again the physicist breaks through the mistaken conventions and liberates reality from misconceptions.

Einstein's loneliness, which so many critics have noticed in this poem, derives partly from the extraordinary sense of space MacLeish creates by portraying the cosmos through Einstein's eyes. He is alone in his contemplation. All other humans are conspicuously absent in this poem as Einstein's isolated intelligence makes "images/ patterned from eddies of the air/ which are/ Perhaps not shadows but the thing itself/ And may be understood."

Still, beyond the power of his prodigious mind, outside the laws of his finite but unbounded universe, Einstein finds *himself* – "Something inviolate. A living something" which cannot be explained by physics and which connects him with the rest of us.

Several other poems by MacLeish explore the concept of relativity. In "Signature for Tempo" the Earth follows an orbital path caused by the space-time curvature. The poet then asks about

> These live people
> These more
> Than three dimensional
> By time protracted edgewise into heretofore
> People,
> How shall we bury all
> These queer-shaped people,
> In graves that have no more
> Than three dimensions?

In denial of an absolute frame of reference, the poem "Immortal Helix" describes the lack of a "real" place for "the fictional" Jacob Schmidt's grave.

> His chronicle is endless – the great curve
> Inscribed in nothing by a point upon
> The spinning surface of a circling sphere.
> Dead bones roll on.

The latter powerful image might have originated in notions of classical physics, where bones seemingly stationary in a grave could be seen as spinning due to the rotation of the Earth. The location of the grave in time and space would then describe not a point, but a circle about the Earth's axis. Adding the motion of the Earth about the sun, the location would describe a great tilted helix, winding along the ecliptic, the Earth's orbital path.

In the context of MacLeish's active interest in Einstein, however, the classical image of a "great curve" inscribed in absolute space changes to the image of a curving of space-time itself, "inscribed in nothing." The notion of a secure resting place is upset much more powerfully by the failure of absolute space itself, than by the mere extension of a point to a helical line.

And a similar concern with "real place and time" appears in "Verses for a Centennial" in which the poet says that Shakespeare's birthplace, or Lincoln's or Dante's,

> Has not been found. They cannot fix their marbles
> Just where the year twelve hundred sixty five
> Rolled up the Arno or where time and Troy

And Stratford crossed each other. On this spot –

> Where now, where now along the great ecliptic
> Traced by a wandering planet that unwinds
> Space into hours?

Our "where" and "when" have no universal meaning; they remain relative to the Earth.

▷   Louis Zukofsky

Like Williams and MacLeish before him, Zukofsky found concepts in the new physics which seemed to offer metaphysical justifications for his experiments in literature. Zukofsky, who translated Anton Reiser's biography of Einstein,[15] said that the aims of poetry "and those of science are not opposed or mutually exclusive; and that only the more complicated, if not finer, tolerance of number, measure and weight that define poetry make it seem imprecise as compared to science, to quick readers of instruments. It should be said rather that the most complicated standards of science – are poetic."[16] Zukofsky elaborates this point by quoting Einstein: "'Everything should be as simple as it can be, but not simpler' – a scientist's defense of art and knowledge – or lightness, completeness and accuracy."[17] Zukofsky wanted the same for poetry. One contemporary instance of this aesthetic is Charles Olson's statement (which he borrowed from Robert Creeley and passed on to William Carlos Williams) that form is never more than an extension of content. In other words, the poet no longer fits his content into already given forms such as sonnets or quatrains, but instead allows the form and content to emerge together as an organic whole.

These innovations in form shocked many who expected traditional structures in poetry and who would agree with Frost that writers experimenting with so-called "free verse" were playing tennis with the net down. Literary critics often describe the contrast between conventional form and more experimental modern structures as the difference between closed and open form. Poems written in closed form refer to governing principles external to the work, principles of meter, rhyme, and stanzaic pattern such as the fourteen line iambic pentameter of the English sonnet with its five cluster rhyme scheme, used in Frost's poem, "Any Size We Please."

Open form, on the other hand, refers to poems whose structures are relative to the content of the poem; as X. J. Kennedy describes it, "Free to use white space for emphasis, able to shorten or lengthen his lines to accommodate whatever he is saying, the poet lets his poem choose its shape as it goes along."[18] Both Williams' "St. Francis Einstein of the Daffodils" and MacLeish's "Einstein" exemplify open form, as do the following examples from Zukofsky's "A": *1-12*.[19]

> Natura Naturata –
> Nature as created.
>   He who creates
>   Is a mode of these inertial systems –
>   The flower – leaf around leaf wrapped
>   around the center leaf,
>   Environs – the sea,
>   The ears, doors;
>   The words –
>   Lost – visible.
>
> Asked Albert who introduced relativity –
> "And what is the formula for success?"
> "X = work, y = play, Z = keep your mouth shut."
> "What about Johann Sebastian? The same formula."

Zukofsky translated the Reiser biography of Einstein during the same years he was writing poems *A-5, 6,* and *7*. Several times he transposed material from the biography to the poetry. In the biography Reiser notes that when a German magazine annoyed Einstein with a questionnaire about Johann Sebastian Bach, Einstein replied with irritation: "In reference to Bach's life and work: listen, play, love, revere, and – keep your mouth shut!"[20] In the part of

Poem *A-6* Zukofsky has used the same anecdote. He sees Einstein as a creator, like Bach, who recognized and translated a universal order into his own discipline. Picking up concepts from both fields, Zukofsky said that poetry should be written as relativized fugues, in an interplay between structure and process.

> I'll tell you.
> About my POETICS –
>
> $\int$ music
>   speech
>
> An integral
> Lower limit speech
> Upper limit music

Here, in *A-12*, the poet has defined poetry as an integral equation from calculus which functions between the limits of music and speech, of sound and story.

> The order that rules music, the same
>   controls the placing of the stars
>   and the feathers in a bird's wing.
> In the middle of harmony
> Most heavenly music
> For the universe is true enough . . . .

> Better a fiddle than geiger?
> With either there is so much in 1
> And in One:
>
> $$\int_{-1}^{1} \qquad \int \begin{array}{l} \text{sound} \\ \text{story – eyes: thing thought . . .} \end{array}$$

> Johann Seb Bach, as he calls him,
> Is present
> His legs in a *gigue*
>   Old French, *to dance (giguer)* or *hop*
> From *gigue* (Teuton *geige* – a fiddle)

The poem creates its own contextual relationships within the limits of simple counting (geiger) and music (fiddle), between the limits of 1 and -1. But the theme is repeated with many subtle variations. The geiger counter warns of radioactivity; the violin reminds one of Einstein; the multi-lingual pun connects Bach in a dance position, the fiddle, and the measurer of cosmic radiation. The same order rules the music, the universe, and the poem. There is an underlying structure. As Einstein commented, "It is impossible for me to say whether Bach or Mozart means more to me . . . . I never like a work if I cannot intuitively grasp its inner unity (architecture)."[21]

Einstein's personal habits were also sources for Zukofsky. In Poem *A-8*, Zukofsky transforms this anecdote from Einstein from the Reiser biography: "Nothing is more foreign to him than elegance or ceremonial garb. In this he agrees with Spinoza, who refused a new coat with these words, 'Will that make me a better man? It would be a bad situation if the bag were better than the meat that's in it.'"[22] In the poem, a figure concerned with his own self-importance says,

> Wherever I sit
> Is the head of the table. Not too
> Near Spinoza refusing a new coat:
> It would be a bad situation
> If the bag were better than the meat in it.
>
> Said Albert – where? – in infinite diapers:
> The bitter and sweet come from the outside,
> The hard from one's own efforts.
>
> For the most part, I do the thing which my
>   own nature
> Drives me to do.
> It is shameful to earn so much respect
>   and love for it.
> I live in that singleness painful in youth,
> but delicious in the years of maturity.

The infinite diapers refer to Einstein's eternal innocence and freshness; the remaining section paraphrases Einstein's often repeated expression of wonder that he should be so honored for doing what came naturally to him.

▷   Charles Olson
    A disciple of Pound, Williams, and Zukofsky, Charles Olson became a central figure for his generation of poets (which included such writers as Robert Creeley, Robert Duncan, Allen Ginsberg, Denise Levertov, Frank O'Hara, and Edward Dorn). When Olson wished to apply non-Euclidean geometries to writing, he saw the "art of space . . .arise from the redefinition of the real, and in that respect,

free, for the first time since Homer, the rigidities of the discrete. What is measure when the universe flips and no part is discrete from another part? .... Art had to invent measure anew."[23]

> The metrical structure of the world is so intimately connected to the inertial structure that the metrical field (art is measure) will of necessity become flexible (what we are finding out these days in painting, writing and music) the moment the inertial field itself is flexible. Which it is, Einstein established, by the phenomena of gravitation and the dependence of the field of inertia on matter.[24]

Written in the late 1950's, Olson's influential essay follows the success of Williams, MacLeish, and Zukofsky in finding in relativity a congenial metaphysics for innovations in poetic form. The muse of Einstein served them explicitly and well.

▶   *More cautious receptions*

▷   Robert Frost

Robert Frost's sonnet, "Any Size We Please," has a similar sense of loneliness in cosmic space to that expressed in much of MacLeish's poetry. This time, however, the loneliness occurs in contemplating a Newtonian universe which extends to infinity, while the curved space of modern physics seems a much more friendly place to be.

> No one was looking at his lonely case;
> So, like a half-mad outpost sentinel,
> Indulging an absurd dramatic spell,
> Albeit not without some shame of face,
> He stretched his arms out to the dark of space
> And held them absolutely parallel
> In infinite appeal. Then saying "Hell,"
> He drew them in for warmth of self-embrace.
> He thought if he could have his space all curved,
> Wrapped in around itself and self-befriended,
> His science needn't get him so unnerved.
> He had been too all out, too much extended.
> He slapped his breast to verify his purse
> And hugged himself for all his universe.[25]

Note that Frost shifts from the Newtonian to the modern universe halfway through the poem at the caesura in line seven; the rhyming in that quatrain of "parallel" with "Hell" refers to the infernal loneliness the persona felt confronting absolutes and infinity. On the other hand, Frost treats a comforting curved space ironically. The persona turns inward and shrinks the curved cosmos down to his own size: "He hugged himself for all his universe," and did not feel secure with any larger conception.

Frost's attitude toward science was ambivalent; it simultaneously fascinated and repulsed him. "He was always deeply absorbed by details concerning any new scientific discovery. At the same time, he could be offended by any cocksure scientific manner which [he felt mocked] poetic and religious concerns for true mysteries."[26] When, for example, Frost heard lectures by Niels Bohr and had dinner with him in Amherst in 1923, the poet was intrigued and full of admiration; but when, a few years later, he had dinner with Robert Millikan and J. B. S. Haldane at California Institute of Technology, the poet was rude and disrespectful, claiming "that their thinking was mere metaphor-making – and poor metaphor making."[27] His differing reactions probably had more to do with the different personalities involved in each occasion than with an inherent inconsistency on Frost's part. His poetry shows both an awareness and a rejection of scientific concepts and their potential uses in poetry. In the poem "Skeptic," for example, he mentions the astronomical evidence of the red shift used to support the Big Bang theory of the origin of the universe, but then says, "I put no faith in the seeming facts of light."[28]

▷   Ezra Pound

Although very different from Frost in most respects, Ezra Pound resembled him in his attacks on science. Pound, a moral absolutist, did not reject Einstein the man or his physical theories, but rather the "socialization" of relativity – "the conception it has given rise to in the minds of laymen and many philosophers alike that 'everything is relative' and there are no absolute standards by which men ought to guide their moral responsibility."[29] Pound felt that Einstein, too, rejected such an unwarranted extension of a purely physical theory into ethics and social behavior. Pound wrote in *Guide to Kulchur* that "Al Einstein scandalized the professing philosophists by saying, with truth, that his theories had no philosophic bearing."[30] On the other hand, Ian F. A. Bell, in his *Critic as Scientist:*

The Modernist Poetics of Ezra Pound, has documented how frequently Pound borrowed terms, images, and analogies from science as he developed his criticism and his literary aesthetics.

▷   T. S. Eliot
Eliot also opposed "scientism" and the extreme positivism that denied any reality other than that which could be empirically verified. He especially objected to the extension of science into religion. In "Thoughts After Lambeth" he complained that Anglican bishops "had bestowed this benediction on our latest popular ramp of best-sellers" – the bishops claiming that relativity as popularized by Jeans, Whitehead, and Eddington "provides a climate more favorable to faith in God than has existed for generations."[31] Eliot noted, "I do not wish to disparage . . .the views set forth by Whitehead, Eddington, and others. But it ought to be made clear that these writers cannot confirm anyone in the faith." He respected "such men as Einstein, Schrödinger, and Planck" who so far had not "written a popular book of peeps into the fairyland of Reality" – who had not, like Eddington and others, gotten "loose into the field of religion."[32] In a note he discusses a London *Times* article on Einstein's acceptance of the new theory of the expanding universe proposed by Edwin Hubble and Richard Tolman. "Our next revelation about the attitude of Science toward Religion will issue, I trust, from Dr. Hubble and Dr. Tolman."[33]

Nevertheless, Eliot did not reject the new physics itself – only its unwarranted extension into other disciplines. Among the major writers, Eliot was one of the most informed about the new physics, having studied and lived with Bertrand Russell, and having read some of the works of Einstein, Whitehead, James Jeans, and Arthur Eddington. Eliot's poetry expresses very well several characteristics of the modern world view induced, in part, by the new physics. The multiple viewpoints of relativity and the corresponding loss of the traditional absolutes have parallels in the fragmentary form and tone of despair in *The Waste Land*.

> London Bridge is falling down falling down
>   falling down . . .
> These fragments I have shored against my ruins[34]

And "The Hollow Men" reflects a shattered, meaningless world – "This broken jaw of our lost kingdoms" where "Lips that would kiss/ Form prayers to broken stone."[35]

The loss of traditional absolutes and the rapidly changing ways of seeing bring both fear and confusion. Still, lilacs do grow "out of the dead land" in the opening of *The Waste Land* even though the loss of the old and birth of the new is "cruel." The poet asks what is coming

> Out of this stony rubbish? Son of man,
> You cannot say, or guess, for you know only
> A heap of broken images, . . .[36]

Here Eliot resembles Yeats wondering what new beast "slouches towards Bethlehem to be born." These writers, mourning the loss of traditional values, face in the opposite direction from writers such as Williams who welcome the new world view. For Eliot, modern science *does* indicate a universe in which truths are relative to the observer's reference frames; therefore, for him, a whole world view governed by such science is fragmentary, frightening, void of meaning. Eliot implies in "The Four Quartets" and in "The Rock" that humanity needs other truths, religious and artistic, which reach beyond the limits of science:

> O weariness of men who turn from God
> To the grandeur of your mind
>   and the glory of your action, . . .
> Dividing the stars into common and preferred,
> Engaged in devising the perfect refrigerator,
> Engaged in working out a rational morality, . . .
>   (from "The Rock"[37])

No wonder such a world ends with a whimper.

▷   E. E. Cummings
        In spite of his avant-garde experiments with form, E. E. Cummings also maintained a romantic hostility to science and scientists. Sometimes his criticism seemed gentle, as in his remark, "I have been found guilty of the misdemeanor known as . . . making light of Einstein."[38] More often, however, he saw science as exploiting, depersonalizing, leveling, and destructive. "So far as I am concerned mystery is the root and blossom of eternal verities while, from a scientific standpoint, eternal verities are nonsense & mystery is something to be abolished at any cost."[39] Cummings' poem "Space being(don't forget to remember)Curved" exemplifies his attitude:

> Space being(don't forget to remember)Curved
> (and that reminds me who said o yes Frost
> Something there is which isn't fond of walls)

an electromagnetic(now I've lost
the)Einstein expanded Newton's law preserved
conTinuum(but we read that beFore)

of Course life being just a Reflex you
know since Everything is Relative or

to sum it All Up god being Dead(not to

mention inTerred)
               LONG LIVE that Upwardlooking
Serene Illustrious and Beatific
Lord of Creation,MAN:
                  at a least crooking
of Whose compassionate digit,earth's most terrific

quadruped swoons into billiardBalls![40]

This fractured and rearranged sonnet echoes Pound's and Eliot's complaint about the careless extension of relativity into standards for human behavior. Note that "god" is in small case, while "Lord of Creation,MAN" receives typographical emphasis. Cummings comments that all these achievements in abstract physical concepts have, in effect, reduced man's humanity. While Michaelangelo's God on the Sistine Chapel ceiling creates man with finger touch, man's "compassionate digit" in Cummings' poem pulls the trigger and kills elephants for their ivory. Man is not a safe or wise god for this world.

The more cautious and negative responses to Einstein and his work on the part of Frost, Pound, Eliot, and Cummings contrast distinctly with the generally positive attitudes with which Williams, MacLeish, Zukofsky, and Olson appear to have regarded their contacts with Einstein and relativity. These differences are certainly due in part to the personalities and outlooks of each of the poets. But these variations also demonstrate that Einstein as a muse could be seen in several distinct incarnations. For Williams and MacLeish, Einstein the revolutionary thinker demonstrated unfettered thought. In addition to direct references to Einstein as an individual, inspirations from Einstein's theory informed experiments in poetic structure by Williams, Zukofsky, and many others. The philosophic extensions of the theory of relativity were themes to be developed for poets such as MacLeish, and were intrusions to be challenged for Eliot and Pound. For Frost, Cummings, and many other poets, Einstein's ideas provided metaphor and imagery. Einstein was thus an important cultural force to be noted by all of these major poets, who made very different choices for the form of their notations.

► *Approaches to relativity in fiction*

Writers of prose fiction displayed an equally excited and wide-ranged interest in Einstein's marvels. From use of the profoundly wrong aphorism "everything is relative," to intricate, accurate analogies, the possibilities of Einstein and his theories as models for subject and form were explored by authors of novels and short stories.

A remarkable early exposition of the possibilities appeared in 1924, with Karel Čapek's novel, *Krakatit* (translated into English the following year).[41] Čapek's awareness of science and technology was indicated by mentions in the novel of the leading scientists of the day, including Einstein, Rutherford, Planck, Bohr, and Millikan. The plot concerns an inventor who has discovered a way to release atomic energy, and who is pursued by those who would use his discovery for their own purposes. The technical details are as accurate as they could be in the early 1920's, and atomic energy is correctly seen as a possibility emerging from the radioactivity work of Becquerel and Rutherford, and not from Einstein's theories – a distinction lacking in much subsequent fiction and popular treatments. Speculations about atomic energy were not new, even in 1924, as will be discussed in Chapter 6. A novelist's use of relativity as metaphor and form, however, may have occurred first in *Krakatit*.

The inventor, Prokop, is torn in the traditional struggle between God and the devil for his soul and his discovery. Prokop's bewilderment, in the literal form of a fever, is described by the first metaphor from relativity:

> Inside his head the blows had become faster and more painful.
>
> It appeared that he was moving with the minimum velocity of light; in some way his heart was compressed. But that was only the Fitzgerald-Lorentz contraction, he explained to himself; soon he would become as flat as a pancake. And suddenly there appeared in front of him countless glass prisms; no, they were only endless, highly polished planes which intersected at sharp angles like models of crystals. He was thrown against the edge of one of them with terrible speed.
>
> . . . .Prokop sobbed with fear. This was Einstein's universe and he must get there before it was too late![42]

Failures of the old absolute references for morality and survival are represented in the novel by the new physics. Towards the end of the novel, however, the struggle for mankind's future is described in Newtonian metaphors of force and motion.[43]

Čapek's experiments were tentative first steps in applying the new physics to literary prose. *Krakatit*, and his later novel *The Absolute at Large*, are still structured in conventional linear narrative, although both novels have surrealistic effects which use science in specific images. New structural devices for prose fiction were being invented, however, and writers soon discovered that Einstein's physics lent itself to these extended formal experiments as well as to the kind of individual metaphors and images that Čapek created. Many of these new inventions were parallel developments to those in relativity theory; others were consciously or unconsciously influenced by popularizations of the new physics. As this study demonstrates, both causal relationships and parallels (not causally related) exist between the new literature and the new science.

The structural device of multiple perspectives in fiction or in painting can correspond, either coincidentally or deliberately, to the postulate in relativity that space and time measurements for one observer will not be the same as for another moving in relation to him. Some of the most astonishing formal innovations in modern literature include the multiple narrators (Joyce's *Ulysses*, Faulkner's *The Sound and the Fury*, Durrell's *The Alexandria Quartet*), plural styles and genres (Dos Passos' *U.S.A.*, Joyce's *Ulysses*, Woolf's *The Waves*), alternative viewpoints added through quotation and allusion (*Ulysses* again, Nabokov's *Lolita*, *Pale Fire* and many other works), the fractured and rearranged syntax of Gertrude Stein, and typographical variations on the page (Dos Passos, Faulkner). What "happens" in the literary work, then, often does not imply any real meaning outside of the several reference frames of the narrators. In poetry, Wallace Stevens offers "Thirteen Ways of Looking at a Blackbird."[44] Each viewpoint is legitimate. The traditional omniscient viewpoint no longer convinces. Faulkner comments, "I think that no one individual can look at truth . . . . You look at it and you see one phase of it . . . . But the truth, I would like to think, comes out, that when the reader has read all these thirteen different ways of looking at the blackbird, the reader has his own fourteenth image of that blackbird which I would like to think is the truth."[45]

Another formal approach in modern literature is to use a single narrator, but to incorporate multiple perspectives by moving him around in space and time, or to have him break temporal and spatial boundaries through techniques of free association, dream, hallucination, or insanity. Kafka's and Ionesco's works come immediately to mind along with the many experiments in literary expressionism and surrealism. Borges, too, works very effectively with dream states. Faulkner again: "When you examine a moment, you will walk around it; you are not satisfied to look at it from just one side."[46]

But multiple perspectives had been in use well before Einstein, and some forms of perspectivism are as old as literature itself. An example in American literature is Nathaniel Hawthorne's *The Scarlet Letter* which has a narrative voice providing commentary based on the individual perspectives of several characters. The famous gold doubloon in Herman Melville's *Moby Dick* is perceived in contrasting aspects by various members of the crew of the *Pequod*. Perspectivism before the 20th century, however, contained two possibilities that are explicitly denied in the version which accompanied Einstein's revolution. Earlier writers could offer the possibility of avoiding error in single-perspective information, and the possibility of ultimate completeness when the information for all perspectives was assembled.

In the typical earlier version of this technique, a single character could always learn individual truth, even if no one observer could see all the truths needed to determine the complete picture. In some novels, the narrator or the reader would further assemble information from many characters, and could thus approach the complete larger truth.

Einstein himself was always confident that absolute truth was indeed attainable. Relativity theory provides for the determination of any one observer's view from any other's, given knowledge of the relative velocities of their space-time frames. Thus the *earlier* perspectivism is a fairly accurate analogy to Einstein's Special Theory of Relativity.

Twentieth-century uses of perspectivism often involve unavoidable distortion, uncertainty, and incompleteness, even as Einstein and his theory are invoked (Lawrence Durrell's work is a prime example, as discussed below). This inaccurate but frequent circumstance is one legacy of the early popularizations of Einstein's theories. Those popularizations unfortunately stressed the relativism of time and space over the more subtle absolutes provided by Einstein's theory.

For many modern authors, attempting to incorporate their understanding of Einstein, temporal and spatial coordinates have no absolute meaning. Recognizing that modern man has no universal frame of reference, Virginia Woolf wrote that authors had to "begin by throwing away" the old traditions. "There is," she said, "no universal scale of values . . . . It is no use going to the guide book; we must consult our own minds."[47] Joseph Frank, in his important essay, "Spatial Form in Modern Literature," points out that the major innovation in both modern poetry and modern fiction is the breaking up of temporal sequences and rearranging them into spatial forms. Juxtaposition of multiple time frames forces the reader to "the simultaneous perception in space of word-groups that have no comprehensible relation to each other when read consecutively in time . . . . Aesthetic form in modern poetry [and in much modern prose] is based on a space-logic that demands a complete reorientation in the reader's attitude towards language."[48]

The new writing required spatial apprehension just as Cubism required a temporal apprehension, both arts incorporating an additional dimension. Innovations in literature duplicated the interaction of time and space in Cubism. Picasso's simultaneous presentation of the side and front view of a face parallels the rearrangement of the time sequence in fiction. The opportunity for comparisons of the new style with the new physics was open, and many authors did try out the possibilities.

▷   Lawrence Durrell

Nowhere is a relativity-inspired style more prominent than in Lawrence Durrell's *Alexandria Quartet*. In his famous note prefacing the second novel of the *Quartet*, Durrell declared:

> Modern literature offers us no Unities, so I have turned to science and am trying to complete a four-decker novel whose form is based on the relativity proposition.
>
> Three sides of space and one of time constitute the soup-mix recipe of a continuum. The four novels follow this pattern.
>
> The three first parts, however, are to be deployed spatially . . . and are not linked in a serial form. They interlap, interweave, in a purely spatial relation. Time is stayed. The fourth part alone will represent time and be a true sequel.[49]

Critics have battled over whether this note is, on the one hand, overbearing, pompous, misleading, and inaccurate, or, on the other hand, a crucial key to both the structure and the themes of the *Quartet*. Durrell was certainly serious enough about the value of modern science as a tool for analyzing the work of other 20th-century writers. His book of criticism, *A Key to Modern British Poetry*, includes a broad introduction to modern physics, and insists on the value of physics for understanding writers including Joyce, Proust, Rilke, and Eliot:

> After all, you might ask, what on earth can the relativity theory have to do with T. S. Eliot's style? I am not suggesting that modern poetry is constructed to illustrate the quantum theory, but I do suggest that it unconsciously reproduces something like the space-time continuum in the way that it uses words and phrases: and the way in which its forms are cyclic rather than extended. Time, both in the novel and in the poem, has taken on a different aspect.[50]

Durrell's free interchange of the terms of relativity theory and quantum theory in this passage and throughout *A Key to Modern British Poetry* indicates his confusion of these two different revolutions in physics. That misunderstanding must be considered in analyzing the role science has played in Durrell's work.

Whether or not Durrell's external comments about physics and literature can be trusted, one can always turn to his novels themselves which surely make an attempt to mirror a relativized world. Events and characters in this tale of love, spying, and betrayal, are seen from several viewpoints, and even those change as observers move around, so that Pursewarden's suicide, for example, receives several different and conflicting explanations, which are never reconciled. The characters, moreover, discuss Einstein's relativity. Pursewarden, in *Balthazar*, claims that "the Relativity proposition was directly responsible for abstract painting, atonal music, and formless (or at any rate cyclic forms in) literature."[51] Justine, in the novel of the same name, says, "If I wrote I would try for a multi-dimensional effect in character, a sort of prism-sightedness. Why should not people show more than one profile at a time?"[52] And in *Clea*, the fourth novel, a posthumous note of Pursewarden's says, "The curvature of space itself would give you a stereoscopic narrative, while human personality seen across a continuum would become prismatic."[53] Simply because writers say they are using relativity, of course, does not mean either that they understand it or that their adaptations of relativity principles succeed artistically.

Every important conclusion drawn by any character in Durrell's *Quartet* is distinctly colored by the character's prejudices and previous knowledge. These predispositions form the framework, analogous to the reference frames of relativity, within which the accuracy of "facts" must be assessed. Different identities for a person standing outside the door of a lovers' trysting place are assigned, first by the lovers, then by a spectator outside, and eventually by the person himself. Here, literal as well as metaphoric perspectives are different. The question is ultimately settled, but other matters, such as Justine's motives for conducting her love affair, are not resolved at the conclusion of the quartet.

The end of the last novel, *Clea*, includes "WORKPOINTS" with suggestions for additional novels, which would present the events of the *Quartet* from additional perspectives. No complete truth can ever be available, although each novel adds to the store of valid interpretations of the events of the story and eliminates some errors of fact. In an interview with Jacob Bronowski, Durrell expressed his belief that inescapable relativity of truth was an accurate reflection of the theory of relativity.[54]

Durrell claims to use the relativism of individual observations as an expression of the ultimate relativism of truth, a conclusion in conflict with both the spirit and letter of Einstein's theory. Durrell's notions are much more properly associated with the later quantum theory, which will be treated in the next chapter. A claim could be made that the novels actually reflect Einstein's work better than Durrell's stated intentions, since many consistent truths about the nature of love and art do emerge despite the relativism of individual observations.

As in most matters of literary taste, judgments on the success of the *Alexandria Quartet* span a broad range of conclusions. A continuing stream of criticism and discussion demonstrates that, at the very least, Durrell has written a fascinating and influential set of novels. It is not fatal to the *Quartet* that what Durrell calls Einstein's relativity is actually a distorted blend of two very different ideas from science. As a theorist for scientists, Einstein offered possible laws of nature, which were to be rigorously tested, and then accepted, modified, or discarded. As a muse for artists and writers, however, Einstein functioned as a provider of potent images and ideas, not a giver of laws. Thus many authors, Durrell included, could produce imaginative work in response to stimulations from this muse, even if they perceived a distorted or confused version of its character.

▷   Vladimir Nabokov

Vladimir Nabokov was asked how he felt about C. P. Snow's complaint that there were two separate cultures, a literary one and a scientific one, that few dared bridge.[55] Nabokov replied:

> I would have compared myself to a Colossus of Rhodes bestriding the gulf between the thermodynamics of Snow and Laurentomania of Leavis, had that gulf not been a mere dimple of a ditch that a small frog could straddle . . . . One of those 'Two Cultures' is really nothing but utilitarian technology; the other is B-grade novels, ideological fiction, popular art. Who cares if there exists a gap between *such* "physics" and *such* "humanities." Science means to me above all natural science. Not the ability to repair a radio set; quite stubby fingers can do that. Apart from this basic consideration I certainly welcome the free interchange of terminology between any branch of science and any raceme of art. There is no science without fancy, and no art without facts . . . .[56]

Nabokov's own art is laced with facts, images, and parodies from science. His first American novel, *Bend Sinister*, uses explicit images from classical astronomy, quantum theory, and relativity. One of his last novels borrows an elaborate image from the popularizations of relativity to make the novel's major point about the confusion of objective reality and art. The image appears in a key scene, early in *Transparent Things*:

> A dingdong bell and a blinking red light at the grade crossing announced an impending event: inexorably the slow barrier came down.
>
> Its brown curtain was only half drawn, disclosing the elegant legs, clad in transparent black, of a female seated inside. We are in a terrific hurry to recapture that moment! The curtain of a sidewalk booth with a kind of piano stool, for the short or tall, and a slot machine enabling one to take one's own snapshot for passport or sport . . . . [Then] a double event happened: the thunder of a nonstop train crashed by, and magnesium lightning flashed from the booth . . . . It had nothing to do with a third simultaneous event next door.[57]

The three simultaneous events are the speeding train passing through the station, the flash of lightning from the coin-operated photo-booth on the platform, and the death of the hero's father of a heart attack. This is a crucial scene in the novel, as several critics have recognized.[58]

The elaborate and apparently irrelevant details of the first two events are not merely attempts to get through a painful description by distractions, however. The non-stop train, an observer on the railway embankment, and the lightning flash are all the elements in Einstein's famous example of the failure of simultaneity, in his own first popularization of relativity:

> Up to now our considerations have been referred to a particular body or reference, which we have styled a "railway embankment." We suppose a very long train travelling along the rails with the constant velocity v . . . . People travelling in this train will with advantage use the train as a rigid reference body (co-ordinate system); they regard all events in reference to the train. Then every event which takes place along the line also takes place at a particular point of the train. Also the definition of simultaneity can be given relative to the train in exactly the same way as with respect to the embankment. As a natural consequence, however, the following question arises:
>
> Are two events (e.g. the two strokes of lightning A and B) which are simultaneous *with reference to the railway embankment* also simultaneous *relatively to the train*? We shall show directly that the answer must be in the negative. [Emphases and parenthetical material Einstein's][59]

Einstein's conclusion is that if the postulates of relativity are accepted, then simultaneity is only a relative observation, not a universal fact. Einstein's example is still used in nearly every popularization of special relativity.

The narrator's assertion that the three events were definitely simultaneous is significant in the special circumstances of this particular image, designed by Einstein to show that simultaneity can *not* be universally defined for events separated in space. In *Transparent Things* narrator Hugh Person has reconstructed his memory in the form of an image from relativity, but one which is inappropriate to his version of what happened. Person's obliviousness to that subtlety presages the central theme of this novel: Person's continuous attempts to make his memory and his physical universe subservient to his art are doomed. Person tries to reimagine and even relive past moments of great emotional power, polishing them (as in the description of his father's death) to suit his artistic sensibility. His perception of his environment is distorted by these attempts to make life imitate art, and the distortions prove fatal.

Many of Nabokov's tragic protagonists have this same fatal mission. Humbert Humbert wants Lolita to realize his vision of a perfect nymphet; Hermann in *Despair* imagines his victim to be his exact double because that would make a symmetric as well as successful murder; Charles Kinbote in *Pale Fire* interprets John Shade's poem to be a disguised celebration of Kinbote's own (imagined?) past life.

Relativity and other images from science often serve as Nabokov's metaphors for fact, literal physical "reality,"[60] while some form of art represents fancy, the imagination's "reality." Both fact and fancy are valid conceptions, and useful tools. The danger comes in attempting to impose one over the other. Only a tolerance of the separate validity of each leads to a survival. In *Pnin*, the title character endures by gaining a bemused acceptance of both. Humbert begins to redeem himself by accepting Lolita as a human existing independently of his dream, and Ada recognizes the impossibility of merging all concepts of time into a single vision from either science or art.

Ada is Nabokov's most ambitious project to illustrate the equal authenticity of physical and imaginative "reality." The novel's hero, Van Veen, is a philosopher and psychologist developing a theory of time. Veen's lectures on the subject, including the extended musings comprising Part Four, include presentations of four distinct conceptions of time: classical physical time, Bergsonian time, Einsteinian time, and Veen's own version of time. Part Four is a parody of a college lecture on these alternative treatments of time:

> Pure Time, Perceptual Time, Tangible Time, Time free of content, context, and running commentary – this is *my* time and theme. All the rest is numerical symbol or some aspect of Space. The texture of Space is not that of Time, and the piebald four-dimensional sport bred by relativists is a quadruped with one leg replaced by the ghost of a leg. My time is also Motionless Time (we shall presently dispose of "flowing" time, water-clock time, water-closet time).[61]
>
> . . .

At this point, I suspect, I should say something about my

attitude to "Relativity." It is not sympathetic. What many cosmogonists tend to accept as an objective truth is really the flaw inherent in mathematics which parades as truth. The body of the astonished person moving in Space is shortened in the direction of motion and shrinks catastrophically as the velocity nears the speed beyond which, by the fiat of a fishy formula, no speed can be. That is his bad luck, not mine – but I sweep away the business of his clock's slowing down . . . . (p. 543)

Van is frustrated in his attempt to out-reason science, and is forced to simply "sweep away" all conceptions of time except his own. Ada herself sympathizes with his effort, but concludes: "I wonder if the attempt to discover those things is worth the stained glass. We can know the time, we can know a time. We can never know Time. Our senses are simply not meant to perceive it. It is like – " (p. 563) and that dash ends Part Four. Alfred Appel suggests that this passage leads us back to the first page of the novel itself,[62] a suggestion supported by Van's description of his own book:

"My aim was to compose a kind of novella in the form of a treatise on the Texture of Time, an investigation of its veily substance, with illustrative metaphors gradually increasing, very gradually building up a logical love story, going from past to present, blossoming as a concrete story, and just as gradually reversing analogies and disintegrating again into bland abstraction" (pp. 562-3).

That is an apt description of Nabokov's use of scientific ideas about time throughout *Ada*.

Each of the visions of time from science so casually dismissed by Van Veen in Part Four has been transformed into an active metaphor in the rest of the novel. Table 1 summarizes the features distinguishing each concept of time. For example, classical and relativistic time are embodied in clockworks, including automobiles (pp. 35, 115), and the clocklike behavior of a character who slows down and loses synchronization in mock-relativistic fashion when he moves (pp. 243, 256). Superposition of instants to represent time flow is presented by flickering film images (pp. 485-95). Differing characteristics of non-clock-like times receive equal metaphoric play (for example, pp. 376, 377, 482, 487, 579). Even Van's lecture dismissing relativity has been given as he drives to a resort, and as he arrives he

Table 1. *Theories of time as they are presented in* Ada

Theory	How changing time is represented	Relation of space to time	Meaning of an *interval* of time	Role of past, present, future
Classical physics	superposition of instants (common analogy: illusion of continuous motion in cinema)	completely separate	absolute, well-defined quantity	absolute, well-defined boundaries
Einstein's relativity	superposition of instants	completely interlinked	no absolute *time* interval, but *space-time* intervals are absolute	well-defined, but vary with observer's reference frame
Bergson's *"durée"*	does not flow, but changes in form and content	completely separate	intervals are subjectively defined, not constant	useful but subjective terms
Van Veen's time	does not flow, never changes	completely separate	interval of time has no meaning, since time does not change	past and present intermixed; future does not exist

discovers: "Today is Monday, July 14, 1922, five-thirteen P.M. by my wrist watch, eleven fifty-two by my car's built-in clock, four-ten by the timepieces in town" (p. 551). Poor Van has become a victim of an exaggerated version of the failure of simultaneity and of the same time dilation he has just swept away.

The most elaborate of the metaphors in *Ada* is the twin motif. Brothers, sisters, names, places, all occur in symmetric, if not literal, twin pairs. The hero and heroine turn out to be brother and sister, with symmetric birthmarks. The twin image, richly used throughout history, is also another key analogy used in many areas of science, including the theory of relativity. A major source of Nabokov's scientific twin images is cited in *Ada*, when Van Veen quotes:

> "Space is a swarming in the eyes, and Time a singing in the ears," says John Shade, a modern poet, as quoted by an invented philosopher ("Martin Gardiner") in *The Ambidextrous Universe*, page 165. [p. 542, parenthetical name and spelling Nabokov's]

This passage is a delightful example of "anti-symmetry." In science, anti-symmetry describes two objects which are identical except for a single opposite characteristic. A positron, for example, is a tiny particle identical to an ordinary electron in every respect except that its electric charge is positive instead of negative. Thus the positron is the "anti-matter" equivalent of the electron. In the anti-world of *Ada*, John Shade is a real modern poet, while "Martin Gardiner" is an invented philosopher. In the mundane world, John Shade is an invented character in a previous Nabokov novel, *Pale Fire*, while Martin Gardner was the real editor of a column in *Scientific American* and is the author of a science popularization, *The Ambidextrous Universe*, which does indeed quote "John Shade."[63] Gardner's treatment of time and space symmetries and anti-symmetries, including the "twin paradox" of relativity theory, was a likely inspiration for *Ada*'s elaborate use of twin images from science.

The "twin paradox" involves twin brothers. One remains on Earth, while the other embarks on a journey in a spaceship. While the brothers are in relative motion, each finds his own time unchanged, but expects his twin's time to be slowed down by relativistic time dilation. The apparent paradox occurs when the travelling brother returns to Earth, and compares his clock and his own biological aging with his twin's. Since each thought the other's time would be running slower, what will they find?

This example was put to Einstein in an attempt to discredit the special theory of relativity as being inconsistent. Even today, some of the subtle aspects of the "twin paradox" are debated in journals of physics education, but the overall resolution of the paradox is clear. The twins do not actually have symmetric time-frames. The twin who remains on Earth stays in a single inertial reference frame throughout. The other must change reference frames at least three times: once to go from the Earth frame to one moving away from the Earth (launching the spaceship), once again to move to a frame moving toward the Earth (turning the spaceship around), and once more to come to a stop on the Earth. Careful study of the effects of changing from one frame to another resolves the paradox. The travelling twin will have aged less than his Earth-bound brother when the two finally meet again and compare clocks.

Less glamorous versions of this experiment have actually been performed. Clocks sent on round-the-world voyages on aircraft or space satellites have been compared with twin clocks left on the ground. The voyaging clocks were found to have lost time in comparison to their Earth-bound twins, just as Einstein predicted.

The twin paradox of relativity is brought up explicitly in the parody of Part Four, where Van Veen attributes it to "Engelwein" (p. 543). Throughout the entire novel, Nabokov's twins, in addition to serving art through their expressions of symmetry, also serve as science analogies. Van and Ada's momentous partings and reunions are tests of the relativistic twin paradox, with Van in particular careful to search for signs of unequal aging.

None of the physics in *Ada* deals with literal experimental science. In every instance, the use of physics is metaphoric and allusive, rather than being a literal examination of the science itself. This manipulation of Einsteinian images, along with images of time from psychology and art, presents Nabokov's fundamental view of the separate but equal validity of art and science as approaches to "reality." Representations of time are presented by manipulating them, comparing their various usefulnesses for different human needs. In physics, Aristotelian time was compared with Newtonian, and then Einsteinian time. In art, Aristotelian unity was challenged by naturalism, stream of consciousness, and other experimental forms. Through his manipulations of each of the conceptions in *Ada*, Nabokov demonstrates that art, unlike science, may still select any version. Fact and fancy stand together, with neither imposing an absolute standard upon the other.

▷   Virginia Woolf
Both the new perspectivism and an enormously increased freedom to manipulate time and spatial coordinates characterize much of the work of Virginia Woolf. At the center of the Bloomsbury group, she recognized the new physical world view and sought to create new forms able to carry the new content, but her work does not contain the explicit references to physics that Durrell and Nabokov chose to use. The Bloomsbury group certainly did discuss the potential uses of the new physics, however. Roger Fry, art critic of the Bloomsbury group, drew on his Cambridge education in the sciences to see numerous relationships between the changes in physics and in all the arts. His frequent "raids across the boundaries" – as Virginia Woolf called his invasions of science and art to explain literature – helped Fry formulate his modern aesthetic theory. Woolf dates her own recognition of the revolution: "In or about December, 1910, human character changed . . . . All human relations have shifted . . . . And when human relations change there is at the same time a change in religion, conduct, politics, and

literature."[64] Her awareness of these shifts came through Roger Fry's introduction of post-impressionism and his discussions of relativity; through Bertrand Russell, who closely followed the developments in physics and wrote extensively about them; through her own knowledge of William James who had anticipated many of the new metaphysical positions of modern physics; through her readings of Proust, who in his reflections of Bergson also prompted manipulation of the sense of time; through Bloomsbury discussions; and through newspapers and journals. In her essay, "How it Strikes a Contemporary" Woolf wrote:

> We are sharply cut off from our predecessors. A shift in the scale – the sudden slip of masses held in position for ages – has shaken the fabric from top to bottom, alienated us from the past and made us perhaps too vividly conscious of the present . . . . New books lure us to read them partly in the hope that they will reflect this rearrangement in our attitude – these scenes, thoughts, and apparently fortuitous groupings of incongruous things which impinge on us with so keen a sense of novelty – and, as literature does, give it back into our keeping, whole and comprehended.[65]

Virginia Woolf did not study modern physics, but she had read James Jeans and was acutely sensitive to the changing world view. She consciously and carefully crafted her best work to "reflect this rearrangement in our attitude." In *The Waves* especially the abstract imagery, the pluralistic points of view, the lack of linear plot, the lack of well-defined characters, the plurally related figures and events, the juxtaposition of antithetical elements, and the form of the novel itself all have parallels with relativity and "wave mechanics" (also called quantum theory) – that second revolution in physics which will be dealt with in the next chapter.

Readers of *The Waves* are first struck by seven points of view, one in the italicized interludes and one for each of the six narrators. None of these viewpoints seems to have any ontological priority, although Bernard certainly has the most words and is responsible for the summing up at the end. The voices of the six figures, moreover, all sound the same. What they say differs, of course, but how they say it seems to be one medium. The common voice holds true whether the speakers are children or adults.

Here, then, is a novel without the expected realistic, naturalistic, or romantic differentiation between characters. Woolf laughed at a

London *Times* review of her book which praised her characters even though she meant to have none. Consistency with relativity would preclude any one-person point of view whether on a Cubist's canvas or in a Woolf novel. For Picasso as well as for Woolf, an object or event seen from several points of view is more real than that described by one person.

In Woolf's space-time world, time cannot be converted to a line and divided into universal pieces or points in time, nor can space be absolutely separated from time. Both William James and Henri Bergson anticipated this aspect of relativity. James described linear or clock time as "intellectualism's attempt to substitute static cuts for units of experienced duration"[66] while Bergson denied the reality of an instant. For him, duration was a continuous emergence of novelty, always becoming; it could never be chopped up into instants that have any real existence. In the modern world view, we live in plural space-time in an ongoing process, not in a determined chronological line divided into individual ticks of some universal clock.

Certainly the figures in *The Waves* do not come across as conventional characters, well-rounded in three spatial dimensions moving through clock time. The six figures all speak in a continuous present tense, while the italicized interludes appear in the past tense, forming a linear background (perhaps historical perspective) for the always-becoming present. This background environment follows the path of the sun across the sky in a linear time framework for the multidimensional worlds of the six figures. Providing each figure with a distinct voice would work against Woolf's purposes. She is not, in this novel, as concerned with her figures as she is with the relationships between them, just as relativity is more concerned with relationships between observations than with the observations themselves. No clear-cut distinctions exist between her figures; they overlap and interrelate in complex patterns, separated and yet together, like intersecting waves in one ocean.

Woolf often paints with relativistic images. Lily, in *To the Lighthouse*, combines the contrasting viewpoints of Mr. and Mrs. Ramsay in her painting – Mrs. Ramsay, near-sighted, subjective, and intuitive; Mr. Ramsay, far-sighted, objective, and intellectual. Lily simultaneously moves through time to Mrs. Ramsay and follows Mr. Ramsay's progress across space to the lighthouse, knowing that she needs "fifty pairs of eyes" in order to juxtapose all the fragments in a meaningful whole. Her painting, which she completes just as Mr. Ramsay and James reach the lighthouse, includes the butterfly colors melting into each other associated with Mrs. Ramsay, as well as the strong linear "bolts of iron" linked to Mr. Ramsay. And the painting mirrors the novel, both in process and finished product.

In *The Waves*, spatial and temporal images often merge. A fused image appears in Rhoda's plea as she sits alone in the classroom. "Look, the loop of the figure is beginning to fill with time; it holds the world in it . . . . The world is entire, and I am outside of it, crying, 'Oh, save me, from being blown forever outside the loop of time!'"[67] For Rhoda, "one moment does not lead to another . . . . I cannot make one moment merge in the next. To me they are all violent, all separate" (p. 265). And she cannot locate herself spatially either. "I have no face," she says often. "I am whirled down caverns, and flap like paper against endless corridors" (p. 266). Rhoda has no sense of belonging to a continuum.

At the other extreme, Bernard sees time in three-dimensions, in a metaphor which rounds Rhoda's chalk loop. "Time lets fall its drop. . . . Time, which is a sunny pasture covered with dancing light, time, which is widespread as a field at midday, becomes pendant" (p. 304). Bernard finds himself most complete in the presence of others. "The private room bores me, also the sky. My being only glitters when all its facets are exposed to many people" (p. 304). "My philosophy, always accumulating, welling up moment by moment, runs like quicksilver a dozen ways at once" (p. 327). For Bernard, life is always a process of multiple becomings, a complex interrelationship which changes as "Time [gives] the arrangement another shake" (p. 365).

In imagery corresponding to exaggerated relativistic effects, both Bernard and Neville describe moving trains which lengthen as they slow down. Einstein's popular example of relativity for a train moving through the station involved the failure of simultaneity (which was used by Nabokov as discussed above), but it also concerned the way in which measurements of length and time would differ from within the train and from on the platform. From the point of view of an observer in the station, a train moving near the speed of light would be shortened, and would then lengthen to its "rest frame" size as it slowed down. Neville comments that "The train slows and lengthens, as we approach London" (p. 224), and Bernard recalls, "The train came in. Lengthening down the platform, the train came to a stop" (p. 364). Jinny, too, describes a complementary effect, the flattened look of hedges as she watches them through the windows of the speeding train.

Much of *The Waves* can be further understood in the light of quantum mechanics, the second great revolution in 20th-century physics. However, one of the best bits of advice for approaching relativity-styled modern literature comes from Neville in *The Waves*, who says, "To read this poem one must have myriad eyes . . . . One must put aside antipathies and jealousies and not interrupt. One must have patience and infinite care . . . . Nothing is to be rejected in fear or horror . . . . There are no commas or semicolons. The lines do not run in convenient lengths. Much is sheer nonsense. One must be skeptical, but throw caution to the winds" (p. 314).

▷   William Faulkner

William Faulkner was one of the first American novelists to experiment with these new forms. Faulkner combined space and time imagery in several novels, particularly in *The Sound and the Fury* where all the major figures have different relationships to time. For Benji, who has no sense of past or future, all time remains in a continuous present. Physical sensations release associated clusters of related events, all of which have sensory immediacy for him. Jason, who reduces life to adding machine calculation and who drives himself towards a fixed purpose, is bound to mechanical clock time. He is defeated by his own inability to adjust to the fluctuating demands of human relationships. Quentin, in trying to escape time into a changeless realm where he and his sister Caddy can be "pure," breaks his watch and tears off the hands. But the ticking pursues him, as do such environmental clocks as church bells, factory whistles, and shadows. As a clock chimes, Quentin muses, "It was a while before the last stroke ceased vibrating. It stayed in the air, more felt than heard, for a long time. Like all bells that ever rang still ringing in the long dying light-rays . . . . Like Father said down the long and lonely light-rays you might see Jesus walking."[68] This passage invokes the possibility in one of Einstein's cosmologies that curved space-time could actually return light to Earth ages after it left, letting people today "see Jesus walking" and hear all the bells that ever rang (technically, one would see the bells but not hear them). The curving light contradicts the mechanical progression of a linear universal time.

Looking in a jeweler's window, Quentin sees watches reading "a dozen different hours and each with the same assertive and contradictory assurance that mine had, without any hands at all . . . . Father said clocks slay time. He said time is dead as long as it is being clicked off by little wheels" (pp. 104-5). None of these times is the real time, but then no absolute time exists. Still trying to escape, Quentin is tricked by his stomach: "Eating the business of eating inside of you space too space and time confused Stomach saying noon brain saying eat oclock" (p. 129). And later he hears a train "dying away, as though it were running through another month or another summer somewhere" (p. 149). These relativistic space-times are not what Quentin wants either; he wants to escape any time, relative to any reference system, into a timeless absolute – a total stasis with his sister Caddy and her lost virginity. "If people could only change one another forever that way merge like flame swirling up for one instant then blown clearly out along the cool eternal dark" (p. 219).

Quentin's father, both in this novel and in *Absalom, Absalom!* stands so aloof from events that he examines them as if he were watching a Greek drama, the actors of "heroic proportions, performing their acts of simple passion and simple violence, impervious to time."[69] But Quentin's only escape from time is suicide. And Sutpen, in *Absalom, Absalom!*, fails in his attempt to impose a rigid, fixed order on Yoknapatawpha County, and to make it permanent. As Faulkner says, "The only alternative to change and process...flux and advancement is death."[70]

The novel's form itself must shape the fluctuating content, the complex interrelationships of varying viewpoints, relative truths, temporal dislocations. Faulkner remarked, "It would be fine if people could write in the old simple clear Hellenic tradition . . . . There was a time for that; the time for that is not now."[71] And so he experiments with the spatialization of time, creating in *As I Lay Dying* a sense of non-motion when time and distance demand motion (the need to get Addie's rotten body into the Jefferson cemetery). Darl says, "It is as though time, no longer running straight before us in a diminishing line, now runs parallel between us like a looping string, the distances being the doubling accretion of the thread and not the interval between . . . . It is as though the space between us were time."[72]

Light in August uses space-time as a means of presenting alienation. Characters separated literally by circumstances are separated metaphorically by virtue of their different space-time reference frames. The Rev. Gail Hightower carries his own rigid internal time: "He knows almost to the second when he should begin to hear it, without recourse to watch or clock. He uses neither, has needed neither for twentyfive years now. He lives dissociated from mechanical time. Yet for that reason he has never lost it."[73] Joe Christmas, whose guardian beat him with specifically clock-like regularity, has understood time only in terms of motion and space: "When he thinks about time, it seems to him now that for thirty years he has lived inside an orderly parade of named and numbered days like fence pickets . . . ." He finally loses time altogether: "Time, the space of light and dark, had long since lost orderliness. It would be either one now, seemingly at an instant, between two movements of the eyelids, without warning" (p. 315). "Uncle Doc" Hines and his wife are displaced from other residents of Mottstown by both time and space: ". . .as though they had been two muskoxen strayed from the north pole, or two homeless and belated beasts from beyond the glacial period" (p. 324). Hines' isolation is manifest in his inability to communicate, "It was that his words, his telling, just did not synchronise . . ." (p. 324).

Slow and quiet Byron Bunch peacefully sees the last of his tormentor, the ever-running Brown, as Brown escapes Mottstown by jumping on a moving train. This final separation is marked not only by relative motion, "passing one another as though on opposite orbits" (p. 417), but also by a metaphoric break in Bunch's space-time:

> He is not thinking at all. It is as though the moving wall of dingy cars were a dyke beyond which the world, time, hope unbelievable and certainly incontrovertible, waited, giving him yet a little more of peace. Anyway, when the last car passes, moving fast now, the world rushes down on him like a flood, a tidal wave.
>
> It is too huge and fast for distance and time; . . . . (pp. 417-18)

Endurance is Faulkner's most treasured quality in a character. So the final lines of *Light in August* belong to Lena Grove, who perseveres. Despite her situation marked by desertion, surrounding violence, and "hope unbelievable," she has wound up with a baby, with Byron Bunch courting her, and with arrival in a new state. Her language blends time and space in simple, delighted acceptance of change: "And it did come and it did suit her. Because she said, 'My, my. A body does get around. Here we aint been coming from Alabama but two months, and now it's already Tennessee'" (p. 480).

Space-time fusions also appear in *Absalom, Absalom!* Sutpen, in his description of the Haitian uprising, describes time as "a space the getting across which did indicate something of leisureliness since time is longer than any distance." On the other hand, his "getting from the fields into the barricaded house seemed to have occurred with a sort of violent abrogation which must have been almost as short as his telling about it – a very condensation of time" (p. 249).

When he describes coming down out of the mountains, Sutpen cannot remember whether the seasons "overtook and passed them on the road, or whether they overtook and passed in slow succession the seasons...they not progressing parallel in time but descending perpendicular" (p. 224). When, in another confrontation with time, Sutpen turns sixty years old, he realizes the biological need for haste and sees time as a thread on a spool *"and that spool almost near enough for him to reach out his hand and touch it"* (p. 279). Time, for him, exists as lines or threads. It becomes a linear segment of his design – the part which determines the schedule and which, therefore, remains as closed and rigid as the design itself. When he calculates that advancing age limits his potency, that the *"old gun...might not [have] enough powder for both a spotting shot and then a full-sized load"* (p. 279), he propositions Miss Rosa. The same inhuman treatment of Milly as a mechanical component in his design leads directly to Sutpen's death. Wash hears the insult and mows Sutpen down with the scythe – symbol of father time and death. His fixed, chronological time table neither controls nor describes the flux of reality around him.

These manipulations by Faulkner of space and time are not explicit experiments with Einstein's space-time, but do offer close parallels. The cited works include many of the same devices that Nabokov's did, reversing space and time images, for example, but without the direct references to science. As with Woolf, Faulkner's space-time style may represent either influence or a deeper parallel evolution of the space-time concept.

▷   Joyce's *Ulysses*
The most influential modern novel is James Joyce's *Ulysses*, which incorporates all the previously discussed innovations from multiple, moving narrators to contortions in time duration and sequence. *Ulysses* does not proceed in a narrative, temporal sequence, but rather by a simultaneous juxtaposition of episodes

demanding that the reader continually fit fragments together and keep allusions in mind until the spatial patterns emerge. Noting the obvious parallels between these stylistic innovations and the explicitly relativity-inspired ones of Durrell and other later writers, several critics have related Joyce's ideas to Einstein's. Because of the importance of *Ulysses* and its critical analysis as models for the study of literature, we shall look closely at both the text and the criticism arguing its connections to science.[74]

In the spring of 1919, while Joyce was living in Zurich on the Universitatstrasse[75] and working on *Ulysses*, Einstein was staying on the Hochstrasse to give a lecture series.[76] The revolutions these two men created were becoming manifest. An experimental confirmation of Einstein's General Theory of Relativity would soon make Einstein the most famous physicist alive. As the publication of *Ulysses* approached, Valery Larbaud could remark in 1921 that Joyce's notoriety had made him as familiar to the literary world as Freud and Einstein were to the scientific world.[77]

The literary innovations of Joyce and the scientific ones of Einstein had even more in common than their European cultural settings and their contemporary development. Both men were concerned with manipulations of time and space, with the relations between subject and observer, and with the role of language in our understanding of the universe. There are also specific invitations, both internal and external to the novel, to consider the relations between *Ulysses* and the scientific revolution of the early 20th century. The Ithaca section, with its explicit science content, was written in 1921,[78] and Einstein's work had been front-page news since November 1919. Not only had Joyce described Ithaca as "a mathematico-astronomico-physico-mechanico-geometrico-chemico sublimation of Bloom and Stephen,"[79] but he also noted that in revising and proofreading Ithaca, "the question of printer's errors is not the chief point. The episode should be read by some person who is a physicist, mathematician and astronomer and a number of other things."[80]

Ulysses includes experiments with time and space, but those experiments alone do not demonstrate that Einstein served as a muse for Joyce. As in the cases of the novels discussed above, a more careful attention to the text and to the author's interests are essential to clarify whether a given experiment is fruitfully examined in the context of relativity or not. In the case of *Ulysses*, many critics have concluded that Einsteinian relativity is indeed a philosophic subject as well as a formal model.

Richard Kain discusses science and relativity occasionally throughout his study, *Fabulous Voyager*.[81] Kain's uses of the term "relativity" are predominantly in his final chapter, but refer in each instance to single measurements of time or space.[82] These have no particular meaning for Einsteinian physics, and are all values in common use throughout 19th-century classical astronomy. Littmann and Schweighauser accurately and thoroughly examine the astronomical terms in *Ulysses*.[83] All the examples are again comfortably 19th century, and Littmann and Schweighauser suggest mostly direct symbolic uses of that science.

Wyndham Lewis, in 1927, directly claimed a relation between *Ulysses* and Einstein, but more a spiritual connection than a scientific one. "This torrent of matter is the Einsteinian flux. Or (equally well) it is the duration-flux of Bergson – that is its philosophic character, at all events."[84] This requisition of a common feature of two very different approaches to time is too broad to provide any test of Einstein's influence.

Avron Fleishman gives a different direction for searching out connections between *Ulysses* and 20th-century science: style.[85] Fleishman reasonably points out that the science content of *Ulysses*, set in 1904, should not directly mention Einstein's relativity, which was first published in 1905. Joyce's fussiness about literal details applies here, too. Although Fleishman gives one example from Ithaca which he claims is clearly intended to refer to Einstein's cosmology,[86] Fleishman's main comments are on the style of the question and answer catechism of Ithaca, and the implications of that style for the value of the scientific approach. Like a catechism, science normally treats the universe by breaking it up into minutely small elements, then limiting consideration even further, to those few elements that readily lend themselves to treatment with available tools. Ithaca's catechism is also like science, says Fleishman, in that it exhibits "no criterion of esthetic taste or human relevance to direct the response."[87]

The highly self-restrictive inquiry of science, like other methods of knowing, has positive but limited value. The major encouragement Fleishman sees Ithaca offering us is the courage of Bloom, and of humanity, who use these limited tools to probe a terrifying universe.

Science as style is also the most convincing aspect of Tindall's treatment. He also sees the style as cold, to "project the inhumanity of science."[88] Edward Watson, too, finds the science in Ithaca directed at demonstrating the objective, impersonal style of science, to be contrasted with the romantic, humanistic approach of other chapters.[89]

None of this critical material makes a case that *Ulysses* has a specific *content* connection with 20th-century science. Fleishman's arguments come closest, showing that "parallax," a concept from classical science, fits in well with the spirit of relativity if not with its literal content. The question of the style of science, rather than its content, bears further examination however.

The loss of religious faith, along with the gains of scientific knowledge, have left Leopold Bloom in a vast universe without teleological motive or moral center. Newtonian, as well as Einsteinian, world views can (and do) serve to illustrate that position in modern literature. Perhaps comfort lies not in the specific physical facts observed by science, but in its style, so evidently successful as Einstein's theories were celebrated in 1919. *Ulysses* examines the subjective human values in scientific style, rather than trying to apply any of Einstein's new findings themselves.

What is the style of science? Scientists have often described their method of inquiry as not only highly self-restricted in subject and technique, but also totally objective, cold, or even inhuman. Lewis, Tindall, Fleishman, and Watson shared this perception. That attitude may have its origin in the rigid style conventionally used in writing about science. That style is based not on a biographical or linear account of the work that was done, but on an account of only those steps producing the results selected for presentation, after the narrator has reached full intellectual and emotional accommodation to his conclusions. Accounts of the human process of reaching accommodation are deleted.[90]

Ulysses illustrates that the style of written science is indeed only a partial depiction of the human process that has occurred. We are led to laugh at the irrelevancy of answering a question about whether water flowed by a detailed account of the entire water-works. The clumsy, pedantic jargon used to describe Bloom's thoughts on being a cuckold appears needless. But these questions and responses serve subjective, human, ends far more than scientific ones. After a day of humiliation, a water tap that works, and works in a manner Bloom can fully comprehend, demonstrates mastery of at least that phase of life. The pain of returning to a marriage bed still marked by a rival's presence can be distanced by concentrating on the technical evidence. Fear of the coldness of the universe is balanced by the comfort and pride of succeeding in the stunning feat of measuring the universal temperature.

The selection of facts presented in *Ulysses*, and in its contemporary science, involves warm and personal choices: which of the innumerable aspects of the universe should be examined; what possible interpretations exist; which of those interpretations should be accepted; what should be done with the results? A cold intellect suffices to catalog answers, but a totally human act was necessary to select the questions.

Who asks the questions in Ithaca? The selection of subjects for inquiry and the mode of response are entirely human choices. Leopold Bloom's accommodation to a universe of bitter facts is made through a dignity and calm he finds in the style of written science. His needs are met with limited but significant success. The science-minded questioner of Ithaca is no less a human than the impressionistic-, romantic-, or dramatic-minded humans who illustrate the limitation and successes of their styles in other parts of *Ulysses*.

Ulysses, then, does not offer parallels to the images and forms of Einstein's relativity. Nevertheless, *Ulysses* does comment on our understanding of the relations between science and literature by illustrating that we can place no more, but perhaps no less, reliance on the particular style of science than we can on our other modes of inquiry. Success in viewing water-works, atoms and stars is only a hint, but a welcome one, that we might just succeed in understanding civilizations, scientists, and novelists.

▷   *Finnegans Wake*

In contrast to *Ulysses*, Joyce's later masterwork, *Finnegans Wake*, is permeated by both the content and the style of Einsteinian physics. Einstein and problems of space and time appear in many guises. Pub-owner H.C. Earwicker (Here Comes Everybody) has twin sons, Shem and Shaun, who carry on all the battles of opposites. In William York Tindall's analysis, "Their contention is that of being with becoming and of ear with eye or of time with space . . . . If Shem is time or ear and Shaun space or eye, H.C.E. is spacetime as his name implies. Earwicker . . . combines 'ear' or time with 'wick' or place. 'Time, please!' in his pub becomes 'Time, place!'"[91] In a Hegelian synthesis Shem and Shaun or time and space have "seemaultaneously sysentangled themselves" and have become "amallgamated" in H.C.E.

Shem, the time-son, sets up a quiz for Shaun who sees things "from the blinkpoint of so eminent a spatialist. From it you will here notice," say Shem, (alias Professor Jones) "that the sophology of Bitchson [Bergson]...borrowed for its nonce ends from the fiery goodmother Miss Fortune (who lost time we had the pleasure we have had our little *recherche* brush with [Proust])...which is in reality only a done by chance ridiculisation of the whoo-whoo and where's hairs theoretics of Winestain [Einstein]."[92] Then, seeing that his explanations "here are probably above your understanding," he tells a story more suited to "muddlecrass pupils" (p. 152).

The parable of The Mookse and The Gripes, which repeats Shem's and Shaun's hostilities, begins with: "Eins within a space and a wearywide space it wast ere whoned a Mookse" (p. 152). That is a relativized echo of the opening lines of *Portrait of the Artist as a Young Man*: "Once upon a time and a very good time it was . . . ." So being "alltolonely" in his "onesomeness," the Mookse-space goes hiking and finds his brother-complement, The Gripes, with whom he immediately argues. "Is this space of our couple of hours too dimensional for you, temporiser?" (p. 154). In spite of their differences, and also because of them, they end up in the same laundry basket in another space-time fusion.

Joyce also plays with the notion that light traveling in curved space-time might eventually return to its source, so that we could see ourselves from behind. "Down the gullies of the eras we may catch ourselves looking forward to what will in no time be staring you larrikins on the post-face in that multimirror megaron of returning-ties, whirled without end to end" (p. 582). The curved space-time also allows Joyce to break with linear time, as many critics have noted, and to adapt curved space-time to the Viconian cycles which help structure *Finnegans Wake*. In other words, Joyce did not limit himself simply to playing verbal games with concepts of modern physics. Clive Hart argues that: "In *Finnegans Wake*, Joyce has pushed relativity to the extreme and made it a basic aesthetic and structural law."[93] And, of course, hundreds of modern writers have been influenced by Joyce whether or not they realized the connections between the innovations in his art and the new ideas in physics.

▶   *More experiments*

The correspondences to relativity, apparent in some of the

works of authors discussed above, also characterize the writing of many contemporary authors. Both Jorge Luis Borges and Vladimir Nabokov, for example, gain multiple perspectives through allusions, quotations, intricate puns, and insertion of other genres into the narratives. Tony Tanner, in *City of Words*, examines their "lexical playfields" which liberate them from "other people's notions as to how one should use language to organize reality."[94] William Gass, Donald Barthelme, Michel Butor, Julio Cortázar, Joyce Carol Oates, John Barth, among many others have experimented with collage-like juxtapositions of fragments. In *Space, Time, and Structure in the Modern Novel*,[95] Sharon Spencer discusses the "mobile" and "stabile" constructs in prose forms in an analogy to the sculpture of Calder and many others. Spencer demonstrates that these forms belong to the same revolution in the means of describing reality, whether the medium be science, steel, or words. Relativistic-like effects have so infused modern literature that the traditional linear narrative told from a single, uncluttered viewpoint now seems exceptional.

Corresponding changes also appear in 20th-century drama. While the successful mechanistic science of the 19th century had obvious connections with realism and naturalism in the theatre, many playwrights soon departed from realism and naturalism by several different paths: a revived romanticism (as in *Cyrano de Bergerac*); a new symbolism in the works of Yeats, Synge, Hauptmann and others; an expressionism in Strindberg and O'Neill; and various experiments with abstractions. One of the more influential innovators, Pirandello, portrayed a world in which reality depended upon the observer. A student of philosophy (with a Ph.D.), Pirandello knew that modern relativistic thinking could influence metaphysics and ethics. His famous *Six Characters in Search of an Author* challenged many theatrical conventions, from the assumption that a character has a knowable identity to the belief that the stage could somehow imitate reality. The unresolved, conflicting perspectives of the stage manager, author, actors, and characters in this play correspond to the multiple right answers in relativity. The truth for an actor is not the same as the truth for the character whose part the actor takes.

The contemporary theater of the absurd developed after World War II had shattered faith in religion, progress, nationalism, and other "isms" which demanded loyalty to a single perspective. Many of the absurd situations in modern plays grow out of plural and conflicting ways of seeing which do not reach any resolution; for instance, the irreducible ambiguities in Beckett's *Waiting for Godot* prevent the audience from selecting any perspective as the right one. The play's concern with time appears in the characters' boredom and anxiety as time passes and yet they cannot recall when or if something happened. Time, in the play, cannot be safely measured. Gone are the secure, mechanical certainties of the Newtonian clockwork universe.

The new perspectives of time, space, and observation traced in this chapter are fascinating examples of the experimentation in literature that paralleled Einstein's development of the Theory of Relativity. It would be presumptuous, however, to attribute all such experimentation to Einstein's influence alone. In earlier chapters, we have examined similar formal changes that were occurring in each of the arts, during and even before Einstein's own views were formed and announced. In some cases, like Durrell and Nabokov, the evidence is clear that they were inspired by reading popularizations of Einstein's work.[96] In other instances, the new perspectives may have been derived from other sources in the arts or sciences, or independently of any specific influence.

Albert Einstein's victory over the high cultural authority of classical science made further challenge to convention seem a more legitimate and promising activity for other potential cultural rebels. Einstein's stature as a muse depends not on his being a sole source, but rather a spectacularly rich source of image and metaphor, and a personal model for the vigorous and fruitful re-examination of profoundly held concepts.

5

The second revolution

But the individual atom is free: it pulsates as it wants, in high or low gear; it decides itself when to absorb and when to radiate energy. There is something to be said for the method employed by male characters in old novels . . . .
from Bend Sinister *by Vladimir Nabokov*

Even as western civilization learned and mislearned about relativity in the 1920's, Einstein and his colleagues were debating a second revolution in physics. Quantum theory was a dramatic and traumatic break with both relativity and classical physics. Not only did this new theory revise the laws of physics, but it changed the form of those laws, dictating that science could not be perfectly predictive, as classical and relativistic physics were, but could only provide probabilities. The philosophic and epistemological implications of quantum theory are far more radical than those of relativity.

Yet the second revolution did not spur headlines or immediate literary investigations of its meaning. No single event, such as the report in 1919 of the eclipse expedition confirming relativity, served to focus attention on the quantum theory. The theory took longer to develop than Einstein's relativity, so that while the first component of the quantum theory was formulated by 1900, comprehensive treatments were developed only in the mid-1920's. The first popularizations of quantum theory followed those of relativity by a decade, so that quantum theory was rarely presented independently of discussions of the much better known relativity revolution. And finally, no one individual served as a personification of quantum theory, as Einstein had for relativity. Work on quantum theory was done simultaneously by so many physicists, including Einstein, that during the early years of the century parallel theories were developed by different individuals, and were given different names. Even today, physicists refer to the physics resulting from the second revolution as either "quantum theory" or "wave mechanics," names revealing different approaches taken by early theorists.

Although its revelation was slower, the quantum theory may have penetrated our culture no less profoundly than the relativity theory. The primary divergence of quantum theory from relativity is that the leading interpretation of quantum theory requires indeterminacy – an irreducible uncertainty – in the best available descriptions of the physical world. This interpretation presents challenges to the basic assumption of cause and effect. As we have seen, many writers have inaccurately ascribed this indeterminacy to relativity, but relativity still describes a completely knowable, causal universe. The popularizations of the late 1920's and 1930's, attempting to explain both revolutions, blurred them together, so that even today many works of literature use relativity and Einstein as images for indeterminacy. Thus familiarity with both quantum theory and its misinterpretations are important to understand the total impact of the quantum theory.

As with relativity, there are a range of possible ways in which the concepts of quantum theory might relate to contemporary literature. Because of the much more gradual and more confused dissemination of quantum theory, however, it is more difficult to determine that theory's literary influences. In this chapter we will outline this second revolution in physics, and examine works of literature which deal with indeterminacy and other potential structural analogues to the themes of quantum theory.

▶ *More trouble with light*

The origins of quantum theory included, as had the origins of relativity theory, attempts to explain properties of light. In the 1890's, all explanations of light began with Maxwell's electromagnetic theory, which seemed the final word on the ultimate nature of light, describing it as an electromagnetic wave.

How was light produced in the first place? Maxwell's theory explained that vibrating electric charges would create waves of electricity and magnetism. This qualitative model for producing light was the basis for trying to understand the production of light in detail. Any hot body, such as a burning coal, should contain vibrating atoms carrying electric charges, and hence should produce light. But what color of light? A lump of coal glows red when first heated. As the temperature rises, the light emitted becomes brighter, and orange and yellow light is added. At even higher temperatures, blue colors are emitted too, and the mix from red to blue gives a "white-hot" overall appearance. Remarkably, a similar pattern occurs for all solids when they are heated. The challenge was to explain this common color/temperature phenomenon (known as "black-body radiation") using the Maxwellian model of vibrating electric charges.

Large assemblies of vibrating atoms could be analyzed using Newtonian mechanics and statistics. When rigorous Newtonian laws of motion were used to predict the detailed vibrations of electric charges in heated bodies, a description of the theoretical frequencies and amplitudes of vibrations was obtained. From these details the colors of light, depending on frequencies of vibration, and the brightnesses, depending on amplitudes, could be predicted from Maxwell's equations. The final result, however, was impossible. All warm bodies were predicted to glow bright, extreme violet. This clearly unrealistic prediction became known as the "ultra-violet catastrophe."

In 1900, the German physicist Max Planck constructed an equation which accurately described the distribution of colors and intensities given off by real glowing bodies. After a few months, Planck had invented a theory to support his equation. Planck accepted the Newtonian description of vibrating electric charges, and the Maxwellian description of electromagnetic waves, light, given off during those vibrations. But Planck added a strange restriction on exchanges of energy between the vibrating charges and the light. Energy could be exchanged, by emission or absorption, only in certain fixed amounts. An electron with any particular frequency of vibration, say f vibrations per second, could emit or absorb light only in amounts of energy that were integer multiples of h times f, where h was a new absolute constant (now known as Planck's constant). Thus energy transfers could take place involving hf, $2hf$, $3hf$, etc., but not with any values in between these. The smallest amount that could be transferred was hf, one *quantum* of energy.

Imagine a child's swing. Every swing assembly has a particular characteristic frequency, exactly as the pendulum of a clock has a precisely defined frequency. Practical experience shows that a swing or pendulum can have any amplitude, from none at all – the swing at rest – to a vigorous arc. The swing can be started with a gentle shove, and additional energy can be added or removed in any desired amount. But now consider a swing which obeyed restrictions like those Planck imposed on the vibrating electrons. No matter how one tried to change the amplitude of motion of the swing, only pushes or pulls that added exactly one or more quanta of energy could affect the motion of the swing. Such a quantum swing would violate common experience and common sense.

But using this restriction, Planck could apply the equations of Newton and Maxwell to generate his empirical formula that correctly described the emission of light. In addition, the number h was so tiny that the quantum of energy for objects like a clock pendulum or a children's swing was extremely tiny. The steps in energy, in other words, would be so small for normal objects that we would never notice the infinitesimal discontinuity in operating a swing or a clock, even if these did in fact obey Planck's rule. For objects as small as atoms, however, the discontinuity would become significant.

Just as equations of relativity approach Newtonian equations for velocities that are small compared to the speed of light, so quantum theory equations approach Newtonian ones for objects that are large compared to atoms. Codified by Niels Bohr as the "Principle of Correspondence," this rule of merging theories aids physicists in the task of creating equations that might describe strange and "impossible" behavior in new realms, such as the very small or the very fast, but would also be accurate for the realm of the ordinary. "The quantum theory would reign supreme on the atomic scale, classical physics on the gross scale of daily experience. Where they overlap, both theories should give the same answers."[1]

Even if Planck's idea accurately described the colors of light from glowing bodies, and even if his equations did so closely approach accepted Newtonian ones for larger energies and objects, the fundamental idea of an irreducible quantum of energy was completely out of keeping with the Newtonian description of energy. Physicists soon realized that the implications of this new restriction on nature were fundamental. What kind of agent could be imagined that prevented an electron in a vibrating atom from acquiring a fraction of a quantum of energy? No mechanical model would behave in this fashion. The very concept of mechanical models for nature, in addition to the Newtonian versions of those models, was being challenged. Planck found that his concept defied

> . . .all attempts to make it fit in with classical theory in any form. So long as this constant could be considered infinitesimal, as when dealing with large energies or long periods of time, everything was in perfect agreement, but in the general case, a rift appeared which became more and more pronounced the weaker and more rapid the oscillations con-

sidered. The failure of all attempts to bridge this gap soon showed that undoubtedly one of two alternatives must obtain. Either the quantum of action was a fictitious quantity . . . or [it] must play a fundamental role in physics, and proclaim itself as something quite new and hitherto unheard of, forcing us to recast our physical ideas, which, since the foundation of the infinitesimal calculus by Leibnitz and Newton, were built on the assumption of continuity of all causal relations.[2]

In 1905, Einstein used the idea of a quantum of energy in describing another puzzling phenomenon, the photo-electric effect. A piece of metal, such as zinc, is electrically isolated by placing it on a glass or a plastic stand. Then the metal plate is given a charge of static electricity, a surplus of electrons. The charge can be detected easily with instruments. Like charges repel each other, and if the metal were not well insulated, the charges would quickly disperse, discharging the plate. On an insulated plate, however, the charges can escape only if they are given additional energy, enough to break away from the metal and cross the metal-air boundary surface.

Shining light on the metal plate can discharge it. That particular interaction between light and electrons is called the photo-electric effect, and today the phenomenon finds many applications ranging from "electric eyes" that control automatic doors to television cameras. At first, this phenomenon seemed in keeping with 19th-century physics. Light, according to Maxwell, is a wave of electricity and magnetism. Light shining on metal means a wave of electricity impinging on the metal, and electric charges on metal will be accelerated back and forth by the wave. If the amplitude of the wave is large enough, the electrons in the metal can acquire enough extra energy from this acceleration to escape. The color of the light, corresponding to frequency of the wave of electricity, will make no difference. Only the amplitude, the brightness of the light, will determine if electrons can gain enough energy to escape, discharging the metal.

Unfortunately, a decade of experiments showed this explanation to be dead wrong. A red light, no matter how bright, would not discharge the metal. Even a beam intense enough to start melting the plate could not knock electrons off. Yet a deep purple light (rich in ultra-violet) could discharge the metal – no matter how dim the light! A bright purple light would do the job faster, but even a very dim purple light was successful where a vastly more intense red light failed completely. This behavior was exactly the reverse of the Maxwellian prediction, which made color irrelevant, and intensity the only determinant of whether a beam of light could discharge the plate.

Einstein's solution was to use energy quanta, but as a way to describe light itself, not merely the amount of energy exchanged between light and vibrating electrons. Suppose, Einstein suggested, we think of light not as a wave at all, but rather as particles of energy (these hypothetical particles of light later become known as *photons*). The more particles that are present, the brighter the light appears. However, the energy in each of these particles of light would depend only on the color of the light, the property that had corresponded to frequency in the wave theory of light. Using Planck's quantum constant, Einstein suggested that the energy E of a particle of light would be $E = hf$. A bright red light would have lots of particles, but because red corresponds to a low frequency, a small f, the energy E of each particle would be small. Thus individual particles of red light would not be able to give an electron a big enough punch to knock it from the metal. Only if several light particles struck an electron repeatedly might that electron escape, but that is statistically unlikely. Particles of ultra-violet light, however, corresponding to much higher frequencies, a large f, have much more energy per particle. Thus a single collision between a particle of violet light and an electron is enough to knock the electron out.

This remarkably simple explanation agreed with the experiments, and continuing investigations confirmed Einstein's hypothesis in great quantitative detail. Einstein's Nobel Prize in 1921 cited his explanation of the photo-electric effect, not relativity, as the unquestionable evidence of his genius.

Einstein's photo-electric effect paper came out in 1905, the same year as the Special Theory of Relativity. Special Relativity was based on the postulate of a constant speed for light, supporting Maxwell's derivation of a single value for the speed of electromagnetic waves, while Einstein's paper on the photo-electric effect seemed to show light was composed of particles, not waves. Einstein was well aware of the contradiction. In his relativity, light was a wave, spread out over space, with definite frequency and indefinite position. In Einstein's light-quantum, light was a particle for which frequency has no meaning and position is a definite, precise characteristic. Young's slit experiments, a century earlier, had convincingly argued that light could not be a particle, since only waves could explain the interference phenomena. Now Einstein was showing that light could not be regarded as a wave in the photo-electric effect, since only particles could explain the effect. Waves and particles were exclusive concepts. Light could not be both, and yet light seemed to behave as particles for some experiments, and as waves for others.

From 1900 to 1920, related quantum concepts were used successfully to explain many of the remaining major puzzles that had troubled Newtonian physics at the end of the 19th century. The change in the temperature of materials as energy is added ("specific heats"), the particular sets of colors given off by hot gases ("line spectra"), the orbits of electrons in an atom, as well as the distribution of light from glowing solids and the photo-electric effect, could each be explained by quantizing some property such as energy or light. But this success extracted an unacceptable price. Light and energy had to be continuous, like waves, for some experiments and theories, but discrete, like particles, for others. This duality was intolerable, for it made physics internally inconsistent.

This strained state of physics was intensified in the early 1920's. Louis de Broglie, a French physicist, proposed in his 1924 doctoral thesis that if light, traditionally thought a wave, could behave like a particle as well, then matter, such as electrons, traditionally thought of as particles, could behave like waves. An experiment followed. A beam of electrons was directed towards a pair of slits, just as Thomas Young had directed a beam of light towards a double slit in one experiment to test the wave nature of light. The slits for electrons had to be much closer together than for light, according to de Broglie's calculations of the "frequency" and "wavelength" that were to be assigned to electrons of given energies. The result was unambiguous. The electrons produced interference, somehow cancelling each other out at just the locations predicted by de Broglie's "matter-wave" theory. Matter was sometimes a mysterious wave. Waves of light were sometimes particles. Physics appeared to have lost consistency.

By 1930, an entirely new physics had emerged, incorporating not only quanta of energy and waves of matter, but a new description of the knowable universe. Comprehensive theories developed separately by Erwin Schrödinger and Werner Heisenberg in the early 1920's, were shown by the early 1930's to be identical in substance, although the two theoreticians had used very different mathematical language. Arguments continue to this day about the proper interpretation of this quantum theory, but the leading interpretation, developed by Niels Bohr and his colleagues and known as the "Copenhagen Interpretation," stands alongside General Relativity in the foundation of contemporary physics.

▶ *The fuzzy universe*

Erwin Schrödinger had produced a new universal equation, as fundamental to physics as Newton's equations of motion or Einstein's equations of relativity. Schrödinger's equation can be applied for any situation to produce a formula, called the "wave function," which describes all aspects of observable physical reality. The wave function itself has no known physical meaning. The function can be manipulated, however, to provide predictions about observations: the probability of finding an electron in a given location at a given time, the most likely energy for an atom, the possible results of measuring the speed and position of a particle. Knowledge is described as the list of possible "states" in which an entity can exist, and the probability of finding that entity in each of its states. Note that the kinds of knowledge which are available are *possibilities* and their *probabilities*, not unique predictions of the results. This marks the great departure of quantum theory, from a world of unambiguous facts to a world of probabilities, lacking demonstrable causality.

Our language is inadequate for this new world view. "Waves" and "particles" are words defined to describe definite, fixed states of existence. An electron, or a beam of light, is now understood to be neither a wave nor a particle, but an entity that can display "wave-like" or "particle-like" features, depending on the circumstances. The definitions of wave and particle, unfortunately, are mutually exclusive and do not permit conceiving of a single entity that can have features of both particle and wave. A tiny particle cannot pass simultaneously through two slits, separated by a distance very large compared to the size of the particle. Yet "particles" can do just that in de Broglie's experiment. A wave cannot be concentrated into so tiny an area that it can give up all its energy to knock a single electron out of a metal plate. Yet light does that in the photo-electric effect, and still passes through two widely separated slits in Young's experiment. New words have not yet been invented to help us grasp the description of nature provided by the quantum theory.

Quantum theory can be interpreted as revealing the unavoidable influence of any act of observation on the subject being observed. The choice of observing apparatus in the preceding example determined whether light was seen as a wave or a particle. An experiment like the double-slit produced evidence about wave-like states of light. The apparatus used to study the photo-electric effect produced evidence about particle-like states. Thus entities cannot be said to have particle or wave properties independently of the observations made to measure those properties. Subject and observer are inseparably linked in quantum theory.

Niels Bohr's 1927 "Principle of Complementarity" summed up this aspect of quantum theory. To obtain the most complete possible description of nature, we are called upon to use two apparently exclusive descriptions. Although they would be inconsistent if applied simultaneously, they are each essential at different times and thus are complementary. The wave-like and particle-like pictures of light should be seen as complements to each other, both required for achieving the full description of the entity called light.

If the world seems solid and well defined in daily life, that is because we normally do not deal with individual electrons or atoms, but with comparatively huge conglomerates. The greater the mass of an entity, the more dominant its particle-like behavior. A speck of dust contains trillions of atoms, and this bulk behaves in an entirely classical particle-like manner.

The dramatic aspect of quantum theory was thus not its immediate practical importance but the challenge the theory presented to testable causality. An archetypal demonstration of classical Newtonian causality would be predicting the path of any particle, like an electron. Given its position and its momentum (momentum is the product of mass and velocity), a classical physicist should be able to predict perfectly the path of an electron in free space.

But according to quantum theory, the electron is not a simple particle, with point mass, whose position and momentum can both be specified exactly. Indeed, position and momentum turn out to be paired characteristics which cannot both be measured precisely at the same time. Werner Heisenberg's 1927 "Uncertainty Principle" (or Indeterminacy Principle) described this aspect of quantum theory. Position and momentum are one of several sets of "conjugate" pairs, and energy and time are another. The act of measuring one property of a conjugate pair severely and unavoidably restricts measurements on the other.

Suppose a physicist sets out to measure both the position and the momentum of an electron. The position might be determined by placing small volumes of a phosphor, like that used inside the face of a television tube, along a path the electron might take. This material will glow when struck by an electron, so the electron's position can be determined by watching for a spot of light from the phosphor. That very detection, however, will change the original momentum of the electron by an unknown amount. For the phosphor to glow, atoms of the phosphor must suffer a collision with the electron and must receive at least one quantum amount of energy from the electron. That is the same quantum of energy that Planck had discovered at the beginning of the quantum theory. Thus this means of measuring the position of the electron requires a collision to take place, and thus will irretrievably destroy the information the physicist sought about the electron's momentum.

Classically, such a clumsy detection technique could, at least in principle, be improved indefinitely to reduce the disturbance of the momentum to a negligible amount. But Heisenberg found that the smaller the uncertainty attained in measuring position, the proportionately larger the uncertainty produced in momentum. This unavoidable uncertainty trade-off cannot be attributed to failings of the particular instruments used, but is due to the nature of measurements in the physical world. Precise determination of both position and momentum is simply not possible. The fuzziness in knowledge is unavoidable because it is a fuzziness inherent in the universe described by these new laws of physics.

With this loss of knowledge the goal of perfect causal prediction must be abandoned even for so simple a situation as determining the path of a single electron. The determinism of the Newtonian clockwork universe was now out of the question for physicists, because the basic knowledge necessary to make complete deterministic predictions did not exist. The very symbol of the atomic age is an atom drawn with electrons inscribing neat, precise elliptical orbits about a nucleus. But orbits imply perfect definition of both position and momentum, violating the Heisenberg Uncertainty Principle. Today, even the single electron of a hydrogen atom should be more accurately drawn as a fuzzy spherical probability cloud.

With quantum theory, several behaviors which were previously regarded as physically impossible became possible. Consider a ball at rest on the bottom of a box which is open at the top. It would be impossible, classically, for that ball to escape by itself. But we have specified both position ("bottom of the box") and momentum ("at rest") as if we could know such parameters with absolute certainty. For objects as large as a ball, the uncertainties are indeed so small that we would not expect ever to see a trapped ball suddenly appear outside the box, rolling away. But if the box is made of electric fields, and the ball is an electron classically trapped within those fields, then the quantum theory uncertainty becomes serious. In the language of the new physics, the electron's wave function penetrates the walls, giving a high probability of finding the electron trapped, and a low but non-zero probability of finding it outside. Thus the electron can sometimes be observed to have escaped through the classically impenetrable barrier. This phenomenon is known as "tunneling," although this word is inadequate because no tunnel under the barrier is imagined to exist. A device known as a "tunnel diode" makes practical use of this property of electrons, and is employed routinely in television sets.

Radioactivity is another example of this classically impossible quantum physics behavior. The constituents of the nuclei of atoms are trapped by electric and nuclear forces, and classically should never be able to escape. But the wave functions of sub-atomic particles (the old language must be used) cannot be perfectly contained by finite forces, so the functions interpenetrate the entire nucleus and spread beyond any "boundary" that could be classically envisioned. For "stable" atoms, like lead, the nucleus is so well bound that we would not expect, on average, to see one disintegrate in the lifetime of the universe to date. But the nuclei of other atoms, such as radium, are less tightly bound, and components of the nucleus can suddenly break free. According to quantum theory, this process is entirely random. An individual nucleus may remain stable for a billion years after its formation, or for only a fraction of a second. There is no way to determine in advance which any particular nucleus will do. The random clicks of a geiger counter near a radium watch dial are the sounds of an acausal realm.

▶   *Einstein leads the resistance*

There was no disputing that the equations of the new quantum physics were extremely accurate in describing a host of phenomena. Resistance to the new physics came over the interpretation of what the theory implied about the nature of physical law.

Historian of science Arthur Miller has described this period in physics as a loss and then a partial recovery of the physicists' ability to *visualize* meaning. Miller sees the emergence of the "Copenhagen interpretation" by Bohr and his colleagues, including the Principle of Complementarity as a key element, as a demonstration of the central role of personal aesthetics in science:

> . . .the path to regaining visualization is characterized by the high drama of the intense personal struggles among the dramatis personae over their choices of the themata in conflict – continuity versus discontinuity, the usefulness of mathematical models versus mechanistic-materialistic models, and whether to maintain causality. These are among the themata that have emerged from Gerald Holton's pioneering historical case studies as having been of great concern to scientists through the ages. Holton refers to them as "thema-antithema couples."[3] His studies reveal that a scientist's criteria for the choice between a thema-antithema couple cannot be reduced either to logic or to a suggestion derived directly from experimental data. Holton's observation and terminology are applicable here because the choice is based upon the individual scientist's aesthetic. What is so fascinating about the genesis of quantum theory is that not only does the personal nature of the struggle between themata emerge from the scientific papers of the period, but the themata clash here as never before in the history of science.[4]

Einstein himself was the first and most persistent aesthetic protester against what he saw as the abandonment of the goal of complete knowledge. "But, surely God does not throw dice in determining how electrons should go!"[5] In a letter to Max Born, Einstein expressed the same concern: "I find the idea quite intolerable that an electron exposed to radiation should choose *of its own free will*, not only its moment to jump off, but also its direction. In that case I would rather be a cobbler, or even an employee in a gaming-house, than a physicist."[6]

For classicists, and in many ways Einstein was at one with them, continuity in nature had meant a rigorous causality. For them, "the evolution of any physical system could be represented by a continuous chain of events causally related."[7] Continuity meant that the position and momentum of an object at one instant determined its position and momentum at the next. It meant that events take place in the same determined way whether or not one is observing them. It meant that inconsistent answers result from inaccuracies or incompleteness of present knowledge, but are not "built in" to the universal system.

Einstein fought for forty years against accepting the new theory as the most complete description possible of the knowable universe. He proposed to his friend Bohr a series of "thought experiments" that might (but did not) reveal inconsistencies in the theory. He proposed critical tests, such as the Einstein-Podolsky-Rosen experiment, that could check for "hidden variables" that might be producing the apparent uncertainties revealed by quantum physics. Practical means for conducting and interpreting that experiment were achieved recently with the development of new instruments and with a mathematical tool, Bell's Theorem.[8] Experiments are being conducted at the present time which, if successful, will resolve some of the basic dilemmas which Einstein proposed and which have remained unanswered.[9]

Einstein also tried to develop alternative theories, including his proposed "Unified Field Theory," which would replace the quantum theory with an expanded version of General Relativity. Although Einstein stimulated much valuable new work with his efforts, none of them dislodged the quantum theory. In the sub-atomic realm, experiment after experiment has confirmed the theory. Although the effects of the "Uncertainty Principle" can only be detected for masses far smaller than a speck of dust, in the realm of the atom the theory reigns without rival. Einstein never tired, however, of repeating his belief that "God does not play dice with the Universe."[10] Einstein's challenge, as well as the quantum theory itself, are concepts that have been spreading into 20th-century culture.

▶   *Quantum theory reaches beyond physics*

Helping to popularize these quantum theory concepts in his *Physics and Philosophy*, Sir James Jeans reiterated the position he had consistently taken since the new physics emerged. "So much of what used to be thought to possess an objective physical existence now proves to consist only of subjective mental constructs . . . . In this progress towards the truth, let us notice that each step was from particles to waves, or from the material to the mental; the final picture consists wholly of waves and its ingredients are wholly mental constructs."[11]

Many physicists did and still do take exception to such statements. Nowhere does Jeans make the distinction that Herman Weyl does between "the objective state of affairs" and the "subjective appearance" of that state.[12] While Bohr proposes that quantum theory blurred the dividing line between subject and object, de Broglie insists that "physics neither abolishes, nor even diminishes, the traditional distinction between subject and object."[13] For de Broglie, the *description* of reality – not reality itself – is dependent on the observer.

But for Jeans and for many others, the new physics justified dropping Cartesian dualism and accepting instead a philosophy in which all paradoxes, such as wave and particle, were simply complementary aspects of the same mental picture.

The apparent subjectivity in modern physics often came across to the lay public in two opposing impressions. On the one hand, this new physics seemed to have restored importance to the individual. Those inclined to rebel against classical insistence on objective distance gladly accepted what they believed to be this scientific justification of subjectivity. On the other hand, the new physics could be seen to argue for an awareness of man's inadequacy, a fundamental limitation on his ability to know what happens.

The well-established, secure classical system which seemed to correspond to common sense ideas of everyday reality, gave way to a more ambiguous new view of man and his world – an open system of paradoxes and uncertainties in which man dealt with fictions rather than absolute truths. Herbert Muller summarizes the revolution in physics as "the triumph of the postulate over the axiom,"[14] or the shift from scientific laws universally accepted as true to scientific statements assumed without proof to be true. He sees three periods in the history of thought: "A Greek period, metaphysical and idealistic, in which emphasis was primarily upon the observer; the scientific period, semi-empirical and materialistic, in which emphasis was primarily on the thing observed; and the period now dawning in which knowledge is a transaction between the observer and the observed."[15]

In his "Dilemma of Determinism," 1884, William James wrote that "the world must not be regarded as a machine" and that the intellectual absolutists, in their blind fear of ambiguity, mistakenly believed that "the minutest dose of disconnectedness of one part with another . . .would ruin everything, and turn this goodly universe into a sort of insane sandheap or nulliverse."[16] Forty years before physics shifted from certainty and determinism to probabilities and indeterminacy, James had been arguing the same cause; and he found much in Henri Bergson's works that agreed with his own.

Bergson's rejection of determinism is inextricably bound with his concept of duration and its role in the ongoing creative evolution. Briefly, his temporalism means indeterminism. Duration, for Bergson, "is by its own nature forever incomplete, being always a *fait accomplissant* and never a *fait accompli*; in other words, it is a *continuous emergence of novelty* and can never be conceived as a mere rearrangement of permanent and pre-existing units. It never barely *is*; it always *becomes*."[17] For Bergson, then, the future can be discussed only in terms of probabilities, and that hesitation ends only when the future becomes present. Hence, present conditions do not absolutely determine future events. The hesitation has nothing to do with man's incomplete knowledge. Present reality would still hesitate before future possibilities if man did not exist.

Bergson's indeterminism also rests on his denial that any meaning can be given to a definite state of the world at a given instant. He saw an "instant" as an arbitrary spatialized point based on the classical mistake of spatializing time – of converting it to a line and then dividing the line into mathematical points in time or instants. Bergson denies the existence of any durationless instants. In so doing, he denies that any instantaneous cut can be made through a system. This fact (as later physically realized in the Uncertainty Principle) limits how much we can know about the state of that system. No instantaneous cut means no mathematically exact description of the position and momentum of a particle; hence no exact determination of its future position and momentum. In this way, Bergson challenged the classical determinism in a manner quite close to that of Heisenberg and Bohr some years later. De Broglie, who freely acknowledges his debt to Bergson, was particularly impressed by Bergson's argument in *Time and Free Will*, summarized in the line: "The effect will no longer be given in the cause. It will reside there only in a state of possibilities."[18]

Whitehead's indeterminism, similar to that of James and Bergson, began with the denial of any real mathematical points or instants, the acceptance of genuine novelties in the on-going process, and the definition of events as "the repositories of possibilities."[19] For him, the "universe [is] always driving on to novelty."[20] Neither Bergson nor Whitehead ever claimed that an event was totally uncaused, but they did say that an event could not be equivalent with, and reducible to, its antecedents.

Bertrand Russell spoke out even more strongly against classical determinism, finding it a product of habitual wishful thinking. In his *ABC of Relativity* he wrote, "The 'law of universal causation' . . . [is an] attempt to bolster up our belief that what has happened before will happen again, which is no better founded than the horse's belief that you will take the turning you usually take."[21] Russell, surprisingly, made the mistake of claiming that the theory of relativity denied cause and effect. He said, for example, that "the word 'effects' belongs to a view of causation which will not fit modern physics, and in particular will not fit relativity."[22] Yet, as Einstein and nearly all physicists saw it, relativity never challenged determinism; and Einstein's split with the Copenhagen school developed mainly because he could not accept the uncertainty and indeterminism that Heisenberg and Bohr argued. The general public, nevertheless, rarely became aware of the details of the Einstein-Bohr debates; hence, it was more likely to accept Russell's misleading statements as well as those of Eddington and Jeans. In any case, Russell, Eddington, Jeans, and Bergson all championed indeterminism and found scientific justification for their positions. Jeans and Bergson, in particular, at once linked indeterminism in the physical world to the individual's free will.

Heisenberg himself was partly responsible for the general misconception that *all* causality disappeared in quantum mechanics. In 1927 he noted that "since all experiments are subjected to the laws of quantum mechanics . . . the invalidity of the law of causality is definitely proved by quantum mechanics."[23] The causality he referred to was the strict, deterministic causality of mechanics – not all kinds of causality. Quantum mechanics, for example, did not refute the principle of sufficient reason. In fact, as Stanley Jaki points out, Heisenberg "and others, who blithely dismissed causality as such on the basis of quantum mechanics, had to resort to a long chain of causal reasoning to prove the absence of mechanical causality in the realm of quantum mechanics. In this connection one cannot help recalling Whitehead's famous remark about those who spend their lives with the purpose of proving that life is purposeless."[24]

In spite of objections from Einstein and Planck, who did not want to give up determinism in nature, and from several other physicists who carefully defined what kind of causality was involved, the impression reaching most people was that causality as such was dead. According to historian Stanley Jaki: "From the speeches of politicians and even some clergymen to the conferences of sociologists and psychologists, the stunning references to the alleged demise of causality were repeated to no end. The imprecise style of many physicists who happened to amplify the subject just kept the dust whirling merrily."[25]

Such misconceptions about causality led immediately to pronouncements that quantum mechanics had given free will back to humanity. Implied, of course, was the equally problematic statement that classical mechanics had deprived humanity of free will. The two scientists most widely read in England both promulgated this notion. James Jeans claimed that while "classical physics seemed to bolt and bar the door leading to any sort of freedom of the will . . .the new physics shows us a universe which . . .might conceivably form a suitable dwelling place for free men."[26] In 1928, Eddington claimed that physics "has withdrawn its moral opposition to free will."[27]

Most scientists, however, thought the extension of physical theory into ethics was entirely unwarranted. Planck, Einstein, Schrödinger, Poincaré, and many others rejected any ethical interpretations of physics. Einstein said that physics dealt with what was, not with what should be, and that value judgments remained outside the domain of the sciences. At Haldane's dinner party in 1921, Einstein assured the Archbishop of Canterbury that physical theory would have no effect on his "morale."[28] Poincaré wrote in 1913 that "ethics and science have their own domains which touch but do not interpenetrate. The one shows us to what goal we should aspire; the other, given the goal, teaches us how to attain it . . . . There can no more be immoral science than there can be scientific morals."[29]

Despite such clear distinctions made by outstanding scientists, Western man continues to find scientific justification for his ethical positions. Part of the reason behind such justifications lies in the long tradition from Descartes through Hobbes and Laplace of "resolving" the apparent contradiction between scientific mechanical determinism and personal freedom. Part comes from an unthinking transference of physical and metaphysical problems into ethics; and part lies in a naive faith in science – well-established by the 19th century –

which tended to take all dilemmas to science and mathematics for solution.

The arguments between physicists also took place on the metaphysical and epistemological level. On one side of the debate were such figures as Bohr, Born, Heisenberg, Pauli, and Dirac. They held that man cannot discover all characteristics of a particle in order to determine its behavior – not because of any human error or ignorance but because the nature of the physical world is such that "we cannot describe what 'happens' between this observation and the next."[30] Heisenberg expresses this idea as follows: "The term 'happens' is restricted to the observation. Now this is a very strange result, since it seems to indicate that observation plays a decisive role in the event and that the reality varies, depending on whether we observe it or not .... We have to remember that what we observe is not nature in itself but nature exposed to our questioning."[31]

On the other side of the argument, Einstein, Planck, de Broglie and Schrödinger did not want to give up determinism, preferring to believe that there was no possibility of "causeless" events on any level and that indeterminism was not a part of nature itself but only a result of man's ignorance. In one of many comments on this question, Einstein said, "Everything is determined, the beginning as well as the end, by forces over which we have no control. It is determined for the insect as well as the star. Human beings, vegetables, or cosmic dust, we all dance to a mysterious tune, intoned in the distance by an invisible piper."[32] These debates often carried over into the public media, as the preceding quote from the *Saturday Evening Post* demonstrates.

In spite of Einstein's protests, however, modern physics has generally followed the Copenhagen interpretation of quantum mechanics, i.e. the interpretation of Bohr, Heisenberg, and their colleagues. Outside physics, the concept of indeterminacy has led to extreme positivist claims that metaphysics itself is dead. Such claims state that since wave mechanics and uncertainty prove that what happens is limited to the observation, it follows that man cannot talk about independently existing reality at all – that man has no knowledge as to the ultimate nature of things.

Quantum mechanics, then, emphasizes discontinuity, indeterminacy, statistical description, probability, and subject-object coupling or overlap. Disintegration, violence, and derangement appear even on the sub-atomic level. For historian of science and physicist Gerald

Holton, "It is as if after a successful search for simplicities and harmonies in science over the last three centuries, the search has turned to a more direct confrontation of complexity and derangement, of sophisticated and astonishing relationships among strangely juxtaposed parts."[33]

▶   *Quantum theory enters literature*

Uncertainty and discontinuity have always had a place in literature, even if only as contrasts for the time-honored virtues of eternal verity and continuity. In much important 20th-century literature, however, indeterminacy and related themes become central to tone and structure, and it will occasionally be possible to document a renewed interest in these ideas as a direct influence from quantum physics. Such clear-cut examples are rare, because as noted at the beginning of this chapter, quantum theory did not enter the general culture with the headlines and drama that characterized the announcements of Einstein's relativity.

Even within science itself, there are other possible origins for the ascendancy of these topics. Darwin introduced randomness in his theory of natural selection, which depended on an ever-present pool of random variations in species from which the selection process could choose. Nineteenth century physics had profited by using the statistics of random deviations for dealing with the behavior of immense collections of identical particles, such as the atoms in a container of gas.[34]

In contrast, indeterminism in quantum theory is elevated to an essential feature, and not merely a convenience as it was in statistical physics, or an expression of unknown factors, as genetic change was for Darwin. The downfall of confidence in strict causality in 20th-century physics grew to be the most serious incursion of physics into the philosophy of modern times. Thus as we examine uncertainty and discontinuity in the literature contemporary with quantum theory, we shall be concerned less with pinpointing immediate influence than with investigating how closely the literary versions of these themes came to parallel the specific style and central philosophical importance that characterize them in the quantum theory.

Perhaps the poet working most directly with quantum theory has been W. H. Auden, who realized and expressed a significant relationship between the uncertainty principle and the role of art. In his essay "The Poet and the City" he points out that "modern science has destroyed our faith in the naive observation of our senses: we cannot, it tells us, ever know what the physical universe is *really* like . . . . This destroys the traditional concept of *art* as *mimesis*, for there is no longer a nature 'out there' to be truly or falsely imitated; all an artist can be *true* to are his subjective sensations and feelings."[35] Auden is not taking off on some romantic tangent here, but is expressing the need for revision of Aristotle's venerable definition of art, because of the drastic change in world view occasioned by modern physics. In his long poem, *New Year Letter*,[36] Auden explores the same aesthetic problem.

> Art in intention is mimesis
> But, realized, the resemblance ceases . . .
> Becomes, though still particular,
> An algebraic formula,
> An abstract model of events
> (pp. 19-20)

In his recognition of indeterminism, he knows

> . . . all our intuitions mock
> The formal logic of the clock.
> All real perception, it would seem,
> Has shifting contours like a dream
> (p. 32)

> Even though we would all
> Prefer our idées fixes to be
> True of a fixed Reality
> (p. 29)

Auden admits that "process change requires both particle and field theory for its full expression . . . . The laws of nature describe statistical regularities and are changing" (p. 99). In the poem it is the devil who says,

> It is Utopia to be dead,
> For only on the Other Side
> Are Absolutes all satisfied
> Where, at the bottom of the graves,
> Low Probability behaves.
> (p. 38)

The devil tempts us with absolutes. But probability does not "behave" in the real world where each man

> Is but a process in a process
> Within a field that never closes;
> As proper people find it strange
> That we are changed by what we change
> (p. 29)
> . . .

> A particle, I must not yield
> To particles who claim the field,
> Nor trust the demagogue who raves,
> A quantum speaking for the waves
> (p. 51)
> . . .

> In labs the puzzled *Kafkas* meet
> The inexplicable defeat.
> The odd behavior of the law,
> The facts that suddenly withdraw . . .
> The Truth where they will be denied
> Permission ever to reside
> (p. 70)

Reality is not out there, or if it is, we cannot talk about it in itself. We can only talk about reality as it responds to our questioning and interaction. The more precise and accurate we try to be, the more nebulous and evasive the "facts" become. We cannot seem to pin down reality in the lab or in the poem. The closest we can come is described by Auden in an analogue of Bohr's complementarity. "The Devil," said Auden, "is the father of Poetry, for poetry might be defined as the clear expression of mixed feelings."[37]

> O how the devil who controls
> The moral asymmetric souls,
> The either-ors, the mongrel halves
> Who find truth in a mirror, laughs . . .
> Just half-truths we can synthesize;
> So, hidden in his hocus-pocus,
> There lies the gift of double focus,
> That magic lamp which looks so dull

And utterly impractical,
Yet, if Aladdin use it right,
Can be a sesame to light.

(pp. 44-45)

So, in spite of the difficulties in being an artist when the traditional conceptions of art as mimesis have been destroyed, the artist can create from the "gift of double focus" works which combine complementary "either-ors" into a meaningful whole. Auden has created an ironic divorce of form and content by writing this poem about modern indeterminism in such a conventional rhyme and meter.

William Carlos Williams offers another early example of themes from quantum physics appropriated for poetic use. Just as the modern conflict between the continuum in classical physics and relativity theory and discontinuity in quantum physics finds a resolution in the principle of complementarity, so the many contraries between universals and particulars, between the "continuous ground" and the discontinuous things become resolved in William Carlos Williams' *Paterson*. Williams constructs *Paterson* out of such dialectical tensions. Wallace Stevens defines Williams' poetry as "the result of the conjunction of the unreal and the real, the sentimental and the anti-poetic, the constant interaction of two opposites."[38] For Williams, the clash between theme/antitheme pairs takes on major importance, informed by Williams' fascination (see Chapter 4) with the new physics.

A dissonance
in the valence of uranium
led to the discovery
Dissonance
(if you are interested)
leads to discovery.[39]

Philosopher Hans Vaihinger expressed the same concept as "Contradictions are not only undeniable but are the very means by which advances have been made."[40] In physics, Bohr often remarked in the context of quantum theory that a profound truth could be recognized by the fact that its opposite was also a profound truth. "Bohr's turn of mind was essentially dialectical, rather than reflective . . . . [His] method of argument shared with the complementarity principle itself the ability to exploit the clash between antithetical positions."[41] Of course, the juxtaposition of complementary opposites is not a new concept generated by quantum theory, but the tendency to think in those terms and to "exploit the clash" of interacting antitheses does characterize both modern physics and much contemporary literature. Williams' turn of mind and methodology shared much with Bohr's, as they had with Einstein's. In a poet's analog of complementarity, Williams says, "Antagonistic cooperation is the key" (*Paterson*, p. 177).

Williams considered *Paterson* as a particular thing and as a field of action. In doing so, he combined discontinuity and continuity in complementary relation. Several pairings in the poem help reinforce this complementarity. Discontinuity appears, for example, in the water

> divided as the dew,
> floating mists, to be rained down

but it comes continuous again as multiple drops are

> regathered into a river that flows
> and encircles (p. 5)

Similar juxtapositions occur between particular individuals and huge crowds; between a living, supple word and the continuous din ("that unmoving roar" in the waterfall-language); between the mature flower and its undifferentiated ground; between the "radiant gist" and the pitchblende – radioactive ore (p. 178).

Other linked contraries appear in the ascents and descents, in the static and the changing, and in the major thematic polarities of male and female. Paterson, the giant-city-poet, is the male complement of the female land – "a woman like a flower . . . . Two women. Three women./ Innumerable women, each like a flower./But/ only one man – like a city" (p. 7). The aim of the poet is to

> bring himself in,
> hold together wives in one wife
> at the same time scatter it,
> the one in all of them (p. 191)

The poet embraces the foulness and immerses himself in the destructive element, in order to find the Beautiful Thing.

Male-female antitheses work throughout the poem. Because the language has failed them, lovers cannot communicate and potential fruition ends as

a bud forever green,
tight curled, upon the pavement, perfect
in juice and substance but divorced,
divorced from its fellows, fallen low (p. 18)

The key word, "divorce," applies to the numerous failures to
recognize complementary relationships. Things or people, divorced
from their antitheme complements, become static – isolated from
life's processes. Because words are divorced from meaning, people
cannot communicate. Sam Patch died in his leap into the Genesee
River because "speech had failed him" (p. 17). Mrs. Cummings
jumped into the Passaic River because "a language (misunderstood)
pouring (misinterpreted) without dignity, without minister [i.e.
without communication with her minister husband], crash[es] upon
a stone ear" (p. 15). Divorce, then, separates the individual from the
language which should connect him with others, from his environment and historical roots, and from the place which is his field of
action.

    – divorced
    from the insistence of place –
    from knowledge,
    from learning – the terms
    foreign, conveying no immediacy, pouring
    down (p. 83)

The poet's job, then, is to reestablish proper complementary
relationships, to marry words with meaning, male with female,
people with place so that "one phrase . . . will lie married beside
another for delight" (p. 140).

In the laboratory or in the poem, Williams recognizes that the more
closely we pay attention to the precise details (which we can do only
by measuring them in various ways) the more we intervene and
interpenetrate, the more we become part of that field. Things in
motion, changing, mutating, cannot be separated from the motion
itself. Yeats asks, how can we tell the dancer from the dance?
Williams ends *Paterson* with the same concept:

    We know nothing and can know nothing
    but
    the dance, to dance to a measure
    contrapuntally,
    Satyrically, the tragic foot (p. 239).

Other poets have explicitly and accurately welcomed aspects of quantum theory, sometimes in their poetry and sometimes in their statements of their aesthetics. Charles Olson, for example, finds the Principle of Complementarity descriptive of reality as he perceives it. "It is rather quantum physics than relativity which will supply a proper evidence here . . . . For example, that light is not only a wave but a corpuscle. Or that electron is not only a corpuscle but a wave . . . . [Poets should be] aware of the complementarity of each of two pairs of how we know and present the real."[42] And Robert Duncan recognizes that "the uncertainty principle in physical measurements that our own science must face [implies] . . . a universe that is not all to be ours!"[43] For these poets, quantum physics supports a belief that the universe cannot be reduced by reason and measurement to a completely knowable reality, even in theory. Duncan points out that the conventional mind thinks of form as something that is imposed, like a mold or template; anything else, for that mind, lacks form – but he finds the open, rhythmic forms of poetry just as legitimate. The determinism of meter and rhyme has been replaced by flexible, organic forms. For Olson, "The structures of the real are flexible, quanta do dissolve in vibrations,"[44] and poetry should have the same plasticity.

▶ *Prose structure*

Prose structure in the 20th century seems filled with examples of indeterminacy and uncertainty, and these forms differ markedly from traditional styles. Chronological patterns in prose literature conventionally have a beginning, middle, and end – a plot in which the action leads to a climax and resolution. These conventional patterns belong with Euclidean geometry in which a line may be divided into a series of points. They also belong with a classical cause-effect determinism in which the conditions of a present state determine those of a subsequent state. Such linear patterns are closed forms. They are teleological and reflect a hierarchically ordered world in which every element contributes to some end.

According to William James, this ordered world is an "all form."[45] In classical physics, this notion might be compared to a universal reference frame, with all its absolutes and true laws. In literature, the closed forms refer to governing principles external to the work of art – principles by which the artist adjusts his content. Convention provides the structure of the house; the artist working in closed forms begins with that given structure and decorates the inside. Closed forms in literature also express the assumption of subject-object dualism, in that the artist remains independent of that which he observes. If he accepts this assumption, the artist reflects or imitates the outside world in his works, which are often judged by how well they function as mirrors.

These traditional plot structures struck the modern writers as highly artificial contrivances dependent on chronological clock time and Aristotelian logic. In her diary Virginia Woolf describes a visit to Thomas Hardy. "'They've changed everything now,' he said. 'We used to think there was a beginning and a middle and an end. We believed in the Aristotelian theory. Now one of those stories came to an end with a woman walking out of the room.' He chuckled."[46]

Rejecting the old ways, modern writers experimented with a variety of new patterns often wholly unrelated to conventional plots. In a letter to the painter Jacques Raverat, Virginia Woolf wrote that writers had to go beyond "the formal railway line of the sentence" and disregard "the falsity of the past . . . . People don't and never did feel or think or dream for a second in that way; but all over the place."[47] As Woolf wrote in her diary, "This appalling narrative business of the realist: getting on from lunch to dinner; it is false, unreal, merely conventional."[48]

As the linear plot broke up, the resolution of critical tensions also disappeared in much modern prose. The story "stopped," but did not "end" as conventional expectations would have preferred. Or, as in *Finnegans Wake*, the last partial sentence of the novel connects with the first partial sentence, and so the cycle goes on forever.[49] Tensions remained unresolved, often hung in complementary suspension. This open ending permits various possibilities for interpretations, because a set of conditions in the fiction do not necessarily lead to a definite conclusion. This openness may also produce a disquieting malaise because of the uncertainty of who's who or what's happening, in Nabokov's "Signs and Symbols," for example, or in Robbe-Grillet's dehumanizations or Sarraute's "Tropisms."

Many modern plays also lack plot in the conventional sense. When staged in Paris, Ionesco's *The Bald Soprano* had no end; the play simply started over again at the beginning. In Samuel Beckett's *Waiting for Godot*, as the character Estragon says, "Nothing happens, nobody comes, nobody goes, it's awful."[50] The tramps wait in uncertainty for someone or something they're unsure of; they cannot recall why or how long they have been waiting and yet they cannot seem to do anything else. Time drags; nothing happens; changes, by which they might measure sequences, seem contradictory or unreal. "Nothing is certain,"[51] says Estragon. Martin Esslin defines Beckett's plays as "polyphonic; they confront their audience with an organized structure of statements and images that interpenetrate each other and that must be apprehended in their totality . . . [and] which gain meaning by their simultaneous interaction."[52] Esslin's "polyphonic structure" resembles Frank's "spatial form" – both indicating a fracturing of linear causal sequences and a rearrangement of the fragments in non-chronological patterns.

Another method of producing a literary indeterminism and uncertainty appears in the expressionistic and surrealistic techniques, which, while they may keep linear causal sequences, so distort the content that they produce nightmarish effects. While the plot structure may remain fairly conventional, in recognizable chronological order, the bizarre distortions of characters and themes work against the normality of the plot, leaving them in unresolved contradiction. One thinks, for instance, of Kafka's "Metamorphosis" and "The Hunger Artist," of Landolfi's "Gogol's Wife," of Cheever's "The Swimmer."

A major artistic problem in the 20th century is that of inventing forms capable of carrying the indeterminisms, ambiguities, pluralities, uncertainties, and antitheses characterizing modern ways of seeing. Many writers use field forms as functional contemporary structures, since the fields can continuously adjust to accommodate discontinuities within their "range."

Essentially these open forms express field relationships in which each individual or event interacts with all other elements within the field. A shift in any one element alters the field so that the whole complex fluctuates. In Faulkner's *Absalom, Absalom!*, for example, as each character adds information and his own perspective to the Sutpen story, he changes the whole field. At no time can the field become fixed, although many figures try unsuccessfully to impose fixity on the process.

In Woolf's *The Waves* the six figures change as they separate and come together. Bernard, on his way to meet his friends at Hampton Court, remarks, "In a moment, when I have joined them, another arrangement will form, another pattern . . . . Already at fifty yards' distance, I feel the order of my being changed."[53] All relationships in the field shift each time one of the figures comes through the swing doors at Percival's farewell party. Rhoda and Jinny, at their coming-out party, both feel the changes in the field every time another person comes into the room. In Rhoda's frame of reference, each newcomer frightens her as the field changes and she cannot find her place in it. "The door opens; the tiger leaps. The door opens; terror rushes in; terror upon terror, pursuing me" (p. 247). Jinny, on the other hand, primed and ready, welcomes the door's opening as her potential lovers keep entering the field. As Woolf commented in her diary, "Never in my life did I attack such a vague and elaborate design. Whenever I make a mark, I have to think of it in relation to a dozen others."[54]

Indeterminism in both quantum physics and in much 20th-century prose has also implied that randomness and chance are built into the fundamental operation of the universe. The Dada movement cheered this new release from determinism. Richard Huelsenbeck describes a Dadaist as "the man of chance . . . . The motley character of the world is welcome to him but no source of surprise."[55] And Hans Arp claims that "Chance opened up perceptions to me, immediate spiritual insights. Intuition led me to revere the law of chance as the highest and deepest of laws."[56] Dadaist poetry, at its most extreme, was composed by cutting words and phrases from a newspaper, putting them in a paper bag, shaking it, withdrawing the words, and writing them down or gluing them to paper in the random order they appeared. Dada artists constructed collages with similar chance juxtapositions of incongruous elements. This intentional nonsense reflected a world which had lost the old meaningful absolutes, where disorder rather than order described reality. As Hugo Ball remarks, Dadaists know "this world of systems has gone to pieces." Dadaist art "is sympathetic because in an age of total disruption it has conserved the will-to-the-image"[57] – it has shaped the chaos in an image, even if the shaping happened to be whimsical or fortuitous.

In Tom Stoppard's 1974 play, *Travesties*, most of the action is under the control of an old man's unreliable memory, and is made even more uncertain by his fantasies and prejudices. Henry Carr tries to remember the Dadaist Tristan Tzara, and his own encounters in Zurich with James Joyce and Lenin. In the old man's reconstruction,

Tzara tells Carr that "clever people try to impose a design on the world and when it goes calamitously wrong they call it fate. In point of fact, everything is Chance, including design."[58] Later, after Tzara has cut up and rearranged a Shakespearian sonnet, he tells Carr that "All poetry is a reshuffling of a pack of picture cards, and all poets are cheats . . . . It comes from the wellspring where my atoms are uniquely organized and my signature is written in the hand of chance."[59]

While the Dada movement lasted only a few years, its mimesis of a meaningless universe operating by chance carried over into existentialism, surrealism, and the literature of the absurd. The fact that physics seemed to present the same picture could reinforce these concepts of uncertainty and indeterminism. Two key existentialist terms, Heidegger's *Das Nichts* and Sartre's *Le Néant*, refer to that Nothingness which follows from a universe with no meaning external to and objectively independent of the individual. Heidegger believes that confrontation with nothingness, akin to facing a meaningless death, produces dread (angst). "The indefiniteness of *what* we dread is not just lack of definition: it represents the essential impossibility of defining the 'what.'"[60] This comes very close to the essential impossibility in quantum mechanics of describing an event exactly. For the quantum physicists, this means working with probabilities rather than certainties; but for those requiring certainties and a meaningful universe independent of man's perceptions, indeterminism and uncertainty have opened up a chaotic void. The old waiter in Hemingway's "A Clean, Well-lighted Place" wants light "and a certain cleanness and order" to keep nothingness away. His prayer expresses an existential confrontation with indeterminism.

> Our nada who art in nada, nada be thy name thy kingdom nada thy will be nada in nada as it is in nada. Give us this nada our daily nada and nada us our nada as we nada our nadas and nada us not into nada but deliver us from nada; pues nada. Hail nothing full of nothing, nothing is with thee.[61]

For Hemingway protagonists, the night produces so many uncertainties and anxieties – a black emptiness lacking definition and order – that they usually sleep during the day or with the lights on. The Hemingway "code hero" builds his own personal value system, knowing that the traditions and old guidelines have collapsed. With no absolute external frame of reference, each individual must invent his own standards.

The loss of absolutes also implies the "death of God." A discontinuous world has no place for an omniscient, omnipotent God, though it may mean for some that individuals have free will. Sartre wrote, "the existentialist . . . finds it extremely embarrassing that God does not exist, for there disappears with Him all possibility of finding values in an intelligible heaven . . . . In other words, there is no determinism – man is free."[62] The existential insistence on each individual's freedom to choose and to be responsible for what he does contradicts both Newtonian mechanics (as applied by some philosophers) and Freudian psychological determinism. Sartre warns that "any man who takes refuge behind the excuse of his passions, or by inventing some deterministic doctrine, is a self-deceiver."[63] Sartre's world, and to some extent the world described by quantum theory, are not determined or controlled by laws of physical nature or by God.

But man has a natural craving for order and meaning. He wants to make sense of the world around him and of his place in it. Albert Camus, in his *Myth of Sisyphus*, describes this search for clarity and certainty in an uncertain universe, and he complains that modern "science that was to teach me everything ends up in a hypothesis, that lucidity founders in metaphor, that uncertainty is resolved in a work of art . . . . This world in itself is not reasonable, that is all that can be said. But what is absurd is the confrontation of this irrational and the wild longing for clarity whose call echoes in the human heart."[64] Once man recognizes the absurdity of life, he can value existence for itself, not in the context of some external determined order. Thus, Sisyphus can be happy. "This universe henceforth without a master seems to him neither sterile nor futile . . . . The struggle itself towards the heights is enough to fill a man's heart."[65]

Much modern literature portrays an anxious search for meaning in an uncertain world where random incongruity and rapid change describe the usual condition. Often writers attempt to give shape to the randomness – one reason that the artist figure himself is so often the protagonist in a contemporary work. Others portray characters frantically trying to cling to old values in a fragmented world where such values no longer apply. Still others accept the absurdity while attempting somehow to live with it. Often the language itself seems to lose meaning, as if it were imitating a meaningless universe. Talk becomes more noise than genuine communication, especially in such modern dramas as Beckett's *Endgame* and *Waiting for Godot*, and in most of Pinter's and Albee's works. Although noise threatened for a while to overcome structure entirely, a wide diversity of literary inventions emerged as writers as varied as Joyce, Woolf, Faulkner, Ionesco, Beckett, and Nabokov began to shape new techniques to contain the randomness. Each came to deal with the same categories of dilemmas faced by the quantum physicists when their linear, unitary structures broke down, and an unavoidable randomness appeared.

▶   *Subject-object blends*

One strategy for defending an acceptance of indeterminacy in literature is for the author to show, as did the physicist, that the observer and the observed are always intermixed beyond the possibility of separation. While subject-object relationship manipulation was also part of the parallel in literature to relativity, there is a major difference by which we can distinguish parallels to quantum theory. In relativity, the particular datum observed depended upon the reference frame, or perspective, of the observer. But that dependence was determined by the relation between the reference frames of the observer and of the object, and not by the act of observation. In quantum theory, each act of observation itself changes the nature of the observed, regardless of the observer's point of view. Subject and object are separable in relativity; they are inextricably joined by every act of observation in quantum theory.

The once objective, detached author now often overlaps or interacts with his material in 20th-century literature. The omniscient viewpoint occurs rarely. First person narration often concerns itself with the process of writing as the artist/protagonist wrestles with his own creative process in the midst of the story itself. One thinks, for instance, of Joyce's Stephen Dedalus, Borges' Pierre Menard, several of Mann's protagonists, Woolf's Bernard. The narrator of Barth's "Life Story" is at once an author, a character in the story, a reader of it. The original fictional reader in Cortázar's "Continuity of Parks" becomes a character in the novel he's reading. Subject-object boundary lines simply break down or lose their clear-cut distinctions.

Literary characters often overlap each other, making it dificult for the reader to discriminate between them Several igures may speak in the same voice, as they do in Woolf's *The Waves*, or they may fuse in a narrator's perception, as several young girls do in Quentin's understanding in *The Sound and the Fury*. Or, like Kafka's Gregor, they may become metaphorical projections of their own unconscious. Physics demonstrated that our senses do not reveal a complete real world, but only a truth about a given act of observation. A literary counterpart finds characters observing what is true only for them and interacting with what they observe in unpredictable ways. Cortázar's "Blow-up"[66] exemplifiies one such interaction which challenges even the camera's objectivity and denies its ability to freeze reality at a moment in time. the realities shift; time refuses to remain arrested – even in the photograph; the observer projects continuous action into the still shot; no clear boundaries separate plural realities.

In the contemporary awareness of subject-object interaction, writers may deny distinctions between art and life instead of claiming that one imitates the other. Sharon Spencer writes in *Space, Time and Structure in the Modern Novel* that "The need to focus the spectator's attention on what is enclosed within the frame is replaced by the need to call his attention to the interconnectedness of all aspects of reality . . . [to] a new alignment of elements . . . . The open structure is a demand for the reader's acute concentration, and sometimes it is a demand, as well, for his active participation in the recreation of the book."[67] The reader's perspective may be the only way to provide a synthesis, overview, or missing links.

Virginia Woolf's *The Waves*, especially, gives a popular version of that postulate in Heisenberg's Uncertainty Principle which says that the manner of observing an object unavoidably changes the object's properties. In her works, adding an observer to the nexus of relationships alters those relationships. Waiting in Hampton Court, the five figures change their pattern as Bernard approaches. In middle age, Jinny must prepare her face in order to observe herself in the mirror. Rhoda dreads the shock of any observer – the change brought on by any encounter. Bernard recognizes his need for an audience in order to create his stories. Whether the figures seek out company or try to avoid it, they all realize that such exposure entails changes of uncertain dimensions.

Of the many scenes in *The Waves* where observation changes what is observed and an introduction of a new element into a field alters the whole field, perhaps the most dramatic focuses on Neville as he impatiently waits for Percival to appear for the farewell dinner.

Arriving early, Neville watches "the door open and shut twenty times already; each time the suspense sharpens . . . . Already the room, with its swing-doors . . . wears the wavering, unreal appearance of a place where one waits expecting something to happen. Things quiver as if not yet in being" (p. 257). As the other five figures come in one by one, they join Neville watching the swing-door as it goes on opening, and the field keeps changing until Percival finally arrives. "I feel at once," says Bernard, "the delicious jostle of confusion of uncertainty, of possibility, of speculation . . . . There is no stability in this world . . . . All is experiment and adventure. We are forever mixing ourselves with unknown quantities" (p. 256).

In Faulkner's *Absalom, Absalom!* the Sutpen story keeps changing as different observers tell it; no one can predict how it will change nor can anyone change it back into its "original" or "true" state. The real Sutpen story, if it ever existed at all, remains forever unknowable. Shreve alone comes close to realizing that all the re-creations of the story are fictions – mental games invented by each teller who thinks his own version is true. Shreve knows he is re-creating a story. "It's better than the theatre, isn't it. It's better than Ben Hur."[68] The two roommates create "out of the rag-tag and bob-ends of old tales and talking, people who perhaps had never existed at all anywhere, who, shadows, were shadows not of flesh and blood which had lived and died but shadows in turn of what were (to one of them at least, to Shreve) shades too" (p. 303). James Guetti says of *Absalom, Absalom!*, "the entire narrative is an hypothesis"[69] – a mental construct or rather a juxtaposition of several mental constructs invented by the narrators as they project themselves into the story they create.

Most modern writers no longer try to portray the old objective truths about an externally existing reality, partly because many believe physics itself has denied we can know them. Much as the physicists link such conjugate variables as particle and wave, position and momentum, discontinuity and continuity in complementary juxtaposition (not a Hegelian synthesis), so many writers, including Joyce, juxtapose thesis and antithesis in a kind of yin and yang complementarity. Instead of trying to make one set of themata absorb or invalidate the other set, Bohr insisted that both sets belong in complementary relationship. In *Finnegans Wake* such conjugate variables appear in the antagonistic twins, Shem and Shaun, and their many counterparts, such as the Gripes and the Mookse, the

Ondt and the Gracehoper, Butt and Taff, etc. The "doubleparalleled twixtytwins" who are "precondamned, two and true" to come together in H.C.E., "the great personage in whom thesis and antithesis are not presented in battle but in harmony, composed to a synthesis. He is 'Hocus Crocus, Esquilocus'."[70] Regarding this complementarity of warring opposites – "talis-qualis" (such as) and "tantum-quantum" (as much as) – Joyce says, "I am working out a quantum theory about it for it is really a most tantumising state of affairs."[71] And later, he says "that tantum ergons irruminate the quantum urge so that eggs is to whey as whay is to zeed . . . . And this is why any simple philadolphus of a fool . . . may be awfully green to one side of him and fruitfully blue on the other" (p. 167) depending on how we look at him or what questions we ask. In physics, double slits give us a wave picture; photo-electric plates give us a particle picture. Both are correct, in spite of the built-in linguistic paradox. Joyce finds in each man a similar complementarity of antagonisms. "These twain are the twins that tick *homo vulgaris* (pp. 418, 419) . . . humble indivisibles in this grand continuum, overlorded by fate and interlarded with accidence" (p. 472) – determinism and indeterminism in the same package.

Within the field of Virginia Woolf's *The Waves*, paired opposites constantly juxtapose in a kind of suspended tension which remains unresolved. These polarities range from major thematic tensions between isolation and communion to minor imagery of light and dark, inside and outside. Every thesis meets its antithesis somewhere in the novel; usually they work in patterns of repetition and variation.

In *The Waves* each figure is both an individual and a part of the continuum, both a particle and a wave. Communion is achieved through the coming together as separate selves, not by the loss of self. Loss of self leads to nothingness, not to integration. Rhoda, for example, who says she has no face, is the figure least able to integrate her being with others. Bernard certainly loses his sense of self as he leans over the gate; but while he can see through the "thick leaves of habit," what he sees is colorless, eclipsed, empty – "the equal and uninteresting landscape" (p. 375). Only with his "self" can he create forms out of formlessness. Whenever Bernard hears someone say, "Behold, this is the truth," he sees something going on which the truth teller has not included. His frequent "Yes, but" recognizes ambiguities and contradictions, all sides of which are true. Bernard's artistic creations depend on the forms he makes out of the dialectical tensions. In a real sense, *The Waves* exemplifies the creative process in the context of the modern world view.

John Graham has indicated how the path of the sun in the italicized interludes traces the form of a huge wave, and many critics have heard the sound of waves through the novel. Wave patterns occur everywhere in this work. The various relationships of the figures form the pattern of the novel.[72] All six move in temporal and spatial juxtapositions which separate and come together in wave patterns. Table 2 indicates two of the many ways these patterns appear.

Wave patterns form from the tensions of opposites which move from theme to antitheme and back to theme. The first three sections, in which the figures gradually gain identity and separateness, precede one section in which they achieve communion, and that pattern repeats itself in sections five through eight. The entire novel builds from the tensions between isolation and integration, discontinuity and continuity. In motion these tensions create wave patterns, an *a-b-a* rhythm. This patterning on three appears clearly even in the first six lines of the children's statements. Each figure in the second triplet repeats and varies what a complementary figure has said in the first triplet. First speaker Bernard sees a ring hanging in a loop of light; fourth speaker Neville sees a globe hanging. Susan, the second speaker, sees slabs of yellow and purple; Jinny, the fifth voice, sees a crimson tassel with gold threads. Rhoda, the third voice, hears birds chirping up and down; Louis, the last, hears a beast's foot stamping and stamping. The first pattern of 1-2-3; 1-2-3 (see, see, hear; see, see, hear) changes in the second set of triplets to a 1-2-3; 3-2-1 pattern with four of the six figures rematched. Bernard's beads of water now repeat Jinny's drops of water, and so on, until all figures overlap and yet each maintains an identity.

Their statements in these first five units all relate to the three senses of sight, hearing, and touch – sometimes mixed in synaesthesia – "Now the cock crows like a spurt of hard, red water in the white tide." Paired oppositions (dark/light, high/low, animate/inanimate, hot/cold, inside/outside, unison/division, open/closed, surface/-depth) never remain static but touch and complement as they repeat each other with variations. Bernard's ring hanging in a loop of light, for example, contrasts with such sharp angular images as Jinny's harsh, short hairs or Neville's jagged knife scraping fish scales. At the same time, however, the loop of light connects with Susan's caterpillar curled in a green ring, with Rhoda's rounds of white china,

Table 2. *Diagrams of The Waves*

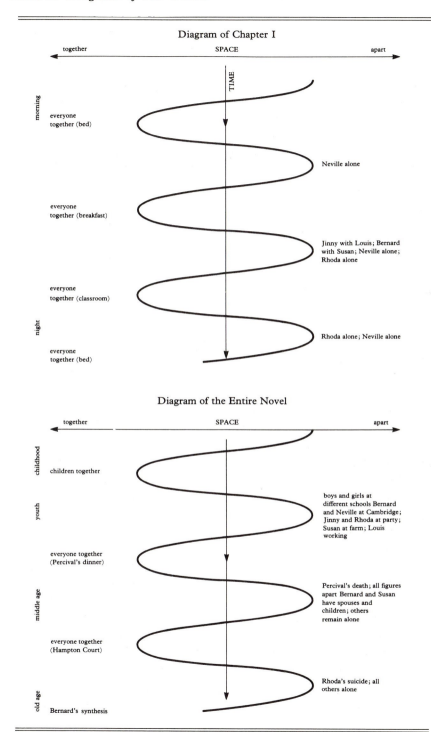

Diagram of Chapter I

together ← SPACE → apart

TIME

morning

everyone together (bed)

Neville alone

everyone together (breakfast)

Jinny with Louis; Bernard with Susan; Neville alone; Rhoda alone

everyone together (classroom)

Rhoda alone; Neville alone

night

everyone together (bed)

Diagram of the Entire Novel

together ← SPACE → apart

childhood

children together

youth

boys and girls at different schools Bernard and Neville at Cambridge; Jinny and Rhoda at party; Susan at farm; Louis working

everyone together (Percival's dinner)

Percival's death; all figures apart Bernard and Susan have spouses and children; others remain alone

middle age

everyone together (Hampton Court)

Rhoda's suicide; all others alone

old age

Bernard's synthesis

with Neville's globe and with numerous drops of water. The children's statements match and contrast simultaneously. This repetition with variation gives the novel its rhythms, its continuously diversifying and unifying formal wave pattern.

Bernard sees the theme-antitheme patterns in complementary relationship. He recognizes the epitomes when all six figures integrate in the brief moments before they, "as a wave breaks, burst asunder" (p. 369). And he knows "the moment was all; the moment was enough" (p. 369). Like waves, the six flow together. "There is no division between me and them" (p. 377). They "rise and fall and fall and rise again" (p. 383) but they also break upon the shore. The final italicized line, *the waves broke upon the shore*, represents Bernard's death or anyone's death, but it also carries the forming template of the next wave. The voices of the six figures emerged from the abstract wave form of the novel, and their interrelationships create the form. The form of the novel is shaped by its content.

Eddington said that the only comprehensive picture of our world given by modern physics is the concept of form. "It is pertinent to remember that the concept of substance has disappeared from fundamental physics; what we ultimately come down to is form . . . . In modern physics, form, particularly wave form, is at the root of everything."[73] Woolf, like many other modern writers, recognized the necessity to create literary form which could carry the concepts of relativity, uncertainty, and complementarity. As she said, "In their shape is their reason."[74]

William Faulkner also expresses the complementarity of conjugate variables and uses the tensions of their paradoxical juxtapositions to help form his fiction. Often the antitheses of determinism and free will appear together in Faulkner's novels in some ambiguous relationship. In *Absalom, Absalom!*, for example, Sutpen consciously refuses to recognize Bon as his son in order to preserve his design. He remains as responsible for that choice as he is for his decision to cast off Milly and her daughter. Faulkner suggests at the same time, however, that events are managed by cosmic forces (through the high-sounding language of several narrators). The time devoted to Sutpen's background also creates a psychological-sociological determinism. Faulkner himself holds an ambiguous position in this regard. "I think that man's free will functions against a Greek background of fate, that he has the free will to choose and the courage, the fortitude to die for his choice . . . That fate – sometimes fate leaves him alone, but he can never depend on that."[75] Part of Quentin's anxiety in his search for understanding of the Sutpen story stems from this complementary tension between free will and determinism, between guilt and innocence.

A few of Faulkner's characters have no free moral choice. In *The Sound and the Fury*, Benji, certainly, cannot be held accountable for his actions; yet he functions as a moral mirror for the actions of others (which Caddy realizes but which Quentin and Jason refuse to see). Figures like Benji, Jim Bond, and Ike Snopes measure the behavior of those characters not trapped from birth in a handicap which prevents their freedom.

Of all the complementary polarities in Faulkner's novels, from oxymorons to major thematic antagonisms, none exceeds the excruciating tension between tragic and comic elements. Faulkner believes "there's not too fine a distinction between humor and tragedy, that even tragedy is in a way walking a tightrope between the ridiculous – between the bizarre and the terrible."[76] In traditional literature, the comic view belongs with a sense of community and human relationships (a continuity), while the tragic view focuses on individuals and broken relationships. In Faulkner's works, the two views almost always appear together – sometimes in outrageous juxtapositions. *As I Lay Dying* is saturated with such tragicomic situations. The confused child, Vardamon, for example, in connecting breathing with life and in trying to comprehend his mother's death, bores holes in her coffin (and into her face) so she can breathe. Anse, trying to set Cash's broken leg, pours concrete on it and then has to chip off the concrete along with a fair amount of skin. In trying to take Addie back to her home town, the family sends townspeople scattering in all directions away from the stench of her rotting corpse. Anse, while he is beholden to no man, depends on everyone else for everything that needs doing – even down to providing the shovels so he can bury his wife. He steals his daughter's abortion money so he can get some new teeth, and he marries, within hours of burying Addie, a woman who has "one of them little graphophones."

In Faulkner's novels as in many other 20th-century works, comedy and tragedy function in complementary juxtaposition, neither one alone being an adequate way of seeing. This is particularly true in the tragicomedy of contemporary drama. Samuel Beckett conveys such juxtapositions in the slapstick comedy/anxiety-ridden tragedy of *Waiting for Godot*. Complementary characters Vladimir and Estragon, and Pozzo and Lucky, wait endlessly in an absurd universe. Beckett's language reflects the lack of meaning. "Where there is no certainty," writes Martin Esslin, "there can be no definite meanings – and the impossibility of ever attaining certainty is one of the main themes of Beckett's plays . . . . In a purposeless world that has lost its ultimate objectives, dialogue, like all action, becomes a mere game to pass the time."[77]

In Harold Pinter's plays, the room inside represents an assumed security while the uncertain, unpredictable menace threatens from outside and usually comes in to change things – Goldberg and McCane in *The Birthday Party*, the matchseller in *A Slight Ache*, Davies in *The Caretaker*, James in *The Collection*, and Ruth in *The Homecoming*. In his plays a mixture of the hilarious and the horrible present the absurdity of the situation. Uncertainty and ambiguity also come from the confused language, misunderstandings, and other breakdowns in communication. Esslin points out that Pinter's dialogue often follows a random associative thinking rather than any logical sequence, reminding one again of *Finnegans Wake* and much of Beckett's work. Both modern physics and modern literature, then, describe the world in terms of uncertainties, polarities, and complementarities.

▶   *Game as metaphor*
The frequent appearance of games, sometimes built in the structure of a work, such as chess in Nabokov's *Lolita*, also corresponds to the physicists' awareness that now they are working with hypotheses, symbolic mappings as Bertrand Russell called them, and not with absolutes. This leaves many writers feeling free to play dice with their fictive worlds.

Robert Coover's novel, *The Universal Baseball Association, Inc. J. Henry Waugh, Prop.*[78] displays an elaborate game world with a chaotic core, and characters who ask what meaning life can have in such a world. A unique feature of Coover's game is the explicit and elaborate construction which propagates the random core into a complete world of everyday events and circumstances. This mechanical device is the most detailed, accurate, and knowledgeable parallel to the quantum theory that we have found in literature, and thus Coover's novel is treated here in appropriate detail. The novel represents not retreat into purely internal reality, but an expansion of

Coover's fiction to include science along with the more traditional sources of metaphor in myth, history, and literature.

In the novel, central character J. Henry Waugh has created the Universal Baseball Association (UBA), a table-top game. The imaginary ballplayers in the UBA have personalities and complete biographies, all established by Waugh's game. The game is played with dice in conjunction with an elaborate set of probability charts. Not only hits, strikes, and balls, but the entire structure of birth, life, and death for each player is determined by the dice and charts. "'There are box scores to be audited, trial balances of averages along the way, seasonal inventories, rewards and punishments to be meted out, life histories to be overseen.' . . . 'People die, you know'" (p. 25).

Although nearly every event in the players' lives is decided by the dice, those events are not strictly arbitrary. The probability charts contain the range of possibilities, and carefully make the occurrence of unlikely events rare. Based on life insurance actuary tables, life, aging and death are eminently reasonable. For example: "Age. It got them all. Began at thirty, a little slower, harder to steal, harder to stretch that long ball into a triple. Injuries tended to be more serious. A little slower afoot out in the field. The slowdown accelerated at thirty-five. All of it worked into the charts" (p. 122). Unusual and even spectacular events are included in the charts, and occur infrequently, lending excitement and life to the game.

Henry's universe, then, is one with rigid, all-encompassing laws (the probability tables), but a core that is random (the dice). The laws can give perfect statistical predictions, yet any single event is completely unpredictable. Over the long run, Henry's players live reasonable lives and, like Henry, wait for those unlikely excursions from the more probable. In Henry's game the random core is only part of the scheme of life. The probability tables are equally crucial and, with the histories and biographies, they determine the limits of the possible, and define the probable.

This scheme is an analogy to the quantum theory's description of reality on an atomic level. The probability tables of Henry's game enumerate the possibilities and assign probabilities to each one by placing every possible event on higher or lower frequency-of-occurrence charts. For example, the "extraordinary occurrences" chart, including death by accident, lists the lowest likelihood events. In a quantum theory description of the energy level changes in an atom, one would find a list of every possible energy shift, and a calculated probability for each. In both Henry's game and the energy transitions of an atom, there is no way to select one event and guarantee that it will happen next. Henry Waugh becomes the god Einstein rejected, one who intends to play dice with his universe.

By the rules of this quantum universe, even a god's knowledge and powers in it are restricted: "Even though he'd set his own rules, his own limits, and though he could change them whenever he wished, nevertheless he and his players were committed to the turns of the mindless and unpredictable – one might even say, irresponsible – dice. That was how it was. He had to accept it, or quit the game altogether" (p. 24).

Henry can only know what the possibilities are, and how often each will probably occur. There is no way for him to know in advance just what will happen. His knowledge is limited in the same way the physicists' knowledge of atomic phenomena is limited. When Henry's favorite player, the young pitcher Damon Rutherford, is struck fatally by a bean ball, Henry can take no action to assure that justice will be done, without destroying the rules of the universe he has set up. "But he sat down anyway, just to see what the dice would bring, because it was clearly on his mind, either something happened – something in short *remedial* – or into the garbage bag with the whole works . . ." (p. 94).

When nothing remedial does happen, Henry, who decides to be a vengeful Old Testament god, must break his own rules and arrange the dice to suit himself, thus forcing the pitcher Jock Casey, who killed Rutherford, also to be slain by a pitched ball. This suspension of the laws of the universe destroys rationality and precipitates Henry's mental disintegration. The price for making the universe inconsistent is insanity.

Henry has been placed in the same dilemma that Einstein envisioned, a god who cannot be content to merely play dice with his universe. Arlen Hansen points out that by naming his players and their teams, by keeping their records, by observing every move of their games, J. Henry Waugh (YHWH – the Hebrew symbols for God) has become personally involved – has lost his objectivity – so that the observer and the things observed interact; the naming of the player changes both the player and the namer. "Name a man and you make him what he is," says Henry as he watches his mental fictions come to life and becomes involved with them. His dilemma mirrors that of the quantum physicist trying to describe quantum behavior in non-mathematical language; as soon as he names an electron a particle or a wave, he creates a false image or at best what Heisenberg calls only a vague connection with reality. If he mistakes his naming of the thing for the thing itself, he prejudices his ability to interpret events.[79]

The situation of the players in the UBA is the same as humanity's in the universe of quantum theory. The players are dimly aware of the true nature of life, and have guessed at the existence of the cosmic probability tables that are nature's only laws: " . . . Barney Bancroft didn't know what Henry knew. He didn't know about the different charts. He didn't even know about Aces and why it was the good ones often stayed good over the years. Of course, he must have sensed it, they all did: that peculiar extra force that these great players seemed to radiate" (p. 34).

One of Henry's creations even guessed at the mathematical details: "Conversations with Fenn McCaffree these days got onesided. He was forever yakking about distribution functions, the canonical form of M, compound decision problems, relations of dominance; like Fenn had somehow forgot the game was baseball" (p. 107).

In such a world, where not even god can have confidence that a system of justice will prevail, the inhabitants wonder if living can have a purpose at all: "Beyond the game, he sees another, and yet another, in endless and hopeless succession. He hits a ground ball to third, is thrown out. Or he beats the throw. What difference, in the terror of eternity, does it make? He stares at the sky, beyond which is more sky, overwhelming in its enormity" (p. 171).

Henry, while still trying to make the universe obey his will and punish Jock Casey, suspects that there might be some *higher* god who can know what the dice will do, and control them. Even an evil god would be better than none at all: "If he didn't know better, he'd suspect the dice of malevolence, rather than mere mindlessness. And it was Henry, not Casey, who was losing control" (p. 111).

Einstein and Waugh both tried to find alternatives to the system, but neither was successful in devising a satisfactory scheme. By the end of the novel, Henry Waugh has disappeared completely, but his universe seems to continue without him. The characters have mythologized those events that occurred when god tried to take a hand, and now must re-examine the structures and myths of their existence.

What sort of hope has been left in a universe without the possibility of deterministic control? Coover gives some choices. First we can appreciate the perfection and balance of the system itself:

> Barney Bancroft had discovered that perfection wasn't a thing, a closed moment, a static fact, but *process*, yes, and the process was transformation, and so Casey had participated in the perfection, too, maybe more than anybody, for even Henry had been affected, and Barney was going to write it . . . And what would Bancroft call it? *The Beginnings*, maybe. Or: *The UBA Story. Abe Flint's Legacy. The UBA in the Balance.* "Yes, that's it" (p. 152).

The random nature of key aspects of the world prevents any higher power from determining our fates, and may still leave some choices open to us, providing the possibility of a limited free will: " . . . the circuit wasn't closed, his or any other; there were patterns, but they were shifting and ambiguous and you had a lot of room inside them" (p. 105).

In our discussions of the parallels between quantum theory and literature, we have found a close similarity in choices of metaphors and themes, but usually without clear cut links that would establish a satisfying cause and effect between this physics and this literature. Coover's novel is a rare example for which there is apparently a direct link. Coover was a philosophy major in college, and when he wrote this novel he was well aware of the history of science, quantum theory, and Einstein's complaint that God does not throw dice.[80] Thus the *Universal Baseball Association* most likely represents a deliberate, accurate, and convincing use in prose fiction of a key philosophical theme from the quantum theory. In a manner appropriate to a chapter on quantum theory, however, the data have been insufficient to describe any general mechanisms through which quantum theory influenced 20th-century literature, as had been possible for the influence of relativity theory. Until an adequate theoretical basis for understanding underlying cultural movement emerges, we must continue collecting data on these fascinating cultural trends, even if, as in the case of the data of life for the UBA teams, apparent realities turn out to be only convincing fictions.

> "It's not a trial," says Damon, glove tucked in his armpit, hands working the new ball. Behind him, he knows, Scat

Batkin, the batter, is moving towards the plate. "It's not even a lesson. It's just what it is." Damon holds the baseball up between them. It is hard and white and alive in the sun (p. 174).

6

A myth portrayed

Just as we have changed our thinking in the world of pure science to embrace newer and more useful concepts, so we must now change our thinking in the world of politics and law. It is too late to make mistakes.
Albert Einstein

Two aspects of the relations between Albert Einstein and the culture of the 20th century have emerged throughout this book. Einstein's revolutionary physics accompanied many parallel dramatic changes in contemporary art and literature. Simultaneously, Einstein as an individual served as a model of the successful challenger to traditional ideas. In this latter connection, Einstein himself appeared as a subject in poems by Williams and MacLeish, and was cited for comparison by numerous poets, novelists, and critics as they embarked on their own challenges to conventional style and structure.

These overwhelmingly positive views of Einstein as an intellectual hero were primarily before the Second World War, however. With new applications of physics to the arts of war between 1939 and 1945 came a major change in how Einstein, the archetype for a physicist, was regarded by writers and by the public. To complete our examination of Einstein's connections to the general culture, we must now look further at the development of Einstein's personal image in the arts and the broader society.

A portrait of extraordinary power appeared on the cover of *Time* magazine for July 1, 1946 (Fig. 3).[1] In the foreground is the face of Albert Einstein, looking very old, with sagging skin, disheveled white hair, and sad eyes. Over his shoulder a mushroom cloud rises on a pillar of multicolored flame, dwarfing a fleet of tiny warships about to be engulfed. Superimposed on the cloud is the equation $E = mc^2$.

This image expresses a common version of the history of nuclear weapons: Albert Einstein, the greatest scientist of the age, created his Theory of Relativity with its central equation, $E = mc^2$, and that equation was the key to the awesome power of the atom. Einstein, the

Fig. 3. *Time* magazine cover for 1 July 1946, by Ernest Tamlin Baker. ©
1946, Time Inc. All rights reserved. Reprinted by permission from *TIME*.

pre-war model of soaring thought for school children and poets, had
inadvertently created the threat of nuclear holocaust which hangs
over mankind.

In present-day discussions of the dilemma of the "atomic age," the
combination of Einstein, the most famous equation, and the mushroom cloud often summarizes the role of science and technology in

society. Great intellectual achievements can make large social changes, but it is argued that those changes are usually unexpected and out of the control even of those who created the ideas. If saintly Einstein was dismayed by this monster of his own making, there is little hope that lesser men, with less humane outlooks, will be able to control the end uses of scientific inquiry.

In a dramatically different fashion then, Einstein re-appeared in post-war newspapers and literature as a tragic figure. In a new version of the myth of Prometheus, Einstein brought the atomic fire to mortal men. This widely used story of Einstein, his equation, and the bomb is nevertheless a gross distortion of the actual history. Serious discussions of atomic energy and atomic weapons had occurred several years *before* Einstein wrote his equation. Einstein's equation played no crucial role in either inspiring or fabricating atomic weapons. The concepts and discoveries which led to the bomb involved a different physics, a different cast of characters, and roughly a half-century of continuous development. Most of that process was deliberate, not inadvertent, and the results, if not the detailed means, had been predicted for forty years.

This chapter will describe the emergence of Albert Einstein's role in a potent contemporary myth, the story of the actual creation of the atomic bomb, and the incorporation of these topics into literature.

▶   *The origin and meaning of $E = mc^2$*
The most famous equation of modern science, $E = mc^2$, was a secondary result of Einstein's 1905 Special Theory of Relativity. The 26-year-old Einstein did not derive $E = mc^2$ as part of his original paper, and the context from which the equation emerged can help to explain why it was never the central expression of relativity that the public today believes it to be.

As discussed in Chapter 3 of this book, Einstein's 1905 theory was concerned with reformulating Isaac Newton's understanding of time and space measurements. In Newton's physics, measurements of length, time, and mass had been assumed to be "absolute," that is, independent of the state of motion of the observer making the measurement. For example, measurements made on a ball flying through the air would give the same mass and dimensions as they would if the ball had been at rest on a table. In contrast, one type of energy possessed by objects was known to be "relative," to depend on any motion of the object relative to the observer. A ball has no "kinetic" energy while it is sitting on the table, but it does have this energy when it is moving. According to Newton, then, mass and dimensions were absolute, while at least one form of energy was relative.

The Special Theory of Relativity reconsidered observations made by observers at different speeds of uniform motion, observers with different "inertial frames of reference." Einstein argued that fundamental properties like spatial dimensions, time, and mass, were in fact all relative to the observer's frame of reference. A ball observed while moving would have not only different energy, but different mass and dimensions as well. The differences in mass and dimension would be tiny under normal circumstances, however, far too small to observe at all unless the ball were traveling at enormous speeds, near the speed of light.

Since both mass and kinetic energy were relative properties, according to Einstein, they might be related to each other in some previously unknown way. The original paper on Special Relativity did not derive any specific relation between mass and energy, although such a relation was implicit in the equations for each one separately. In a three-page paper published later in 1905[2], Einstein did ask about the relations between mass and any form of energy. He considered what would happen when any object, even at rest, changed its energy content. An atom, for example, can store electric energy and then release some of its stored energy by emitting a burst of radiation. In a neon sign, the atoms of neon gas are given extra electric energy by the electric current surging through the tube. Atoms which are "excited" by the current then release that extra energy by giving off radiation in the form of the characteristic red-orange light of neon signs.

Using Special Relativity, Einstein deduced the surprising result that a change in an atom's energy, E, was accompanied by a change in its mass, m. The two properties were related by the equation $m = E/c^2$, where c^2 is a constant number numerically equal to the speed of light multiplied by itself. Neon atoms gain a little mass when they gain electric energy, and then lose that extra mass when they give off the energy in the form of light. Since Einstein's example dealt with light carrying off excess energy, the appearance of the speed of light in the equation was natural.

Two years later, in 1907, Einstein suggested the validity of this equation as a relationship between *all* masses and their total energy content, even if neither light nor motion were involved.[3] He also wrote the equation in a typographically different but mathematically identical form: $E = mc^2$. Any change in one property meant a change in the other. Energy and mass were just different aspects of the same fundamental property of a body. The equation applies not only to atomic energy but to *all* forms, including electrical, thermal, or mechanical energy. If you depress the top of a ballpoint pen, compressing the spring to expose the point, you have added energy to the pen. The pen thus has not only more energy, but also slightly more mass than before, according to $E = mc^2$. Another way you could also increase the mass of the pen would be by throwing it across the room. Its added energy, this time in the form of motion, would increase its mass. Even when it crashed into the wall and came to rest on the floor, it would still have some extra mass, because of the energy in the form of heat that would have been generated during the crash. Although any of these increases in mass would in practice be far too small to detect, they are all perfectly accurate examples of $E = mc^2$.

This equation represents one relationship derived from Einstein's fundamental reconsideration of the relations between measurements of time and space. That derivative equation does not by itself report the full significance of the Special Theory of Relativity, nor does it lend itself to direct inspiration of new forms of literature, as the relativity of time did. The equation carried a special significance for physicists, however, because of the possibility of practical tests of its validity. While all of the other equations of special relativity differ from classical physics only for phenomena near the speed of light, this equation could be tested on *any* phenomenon which involved a large percentage change in total energy. The equation also had potential utility as a measuring tool in nuclear physics, since energy could be deduced from measurements of mass, and vice-versa.

The popular elevation of $E = mc^2$ to represent all of relativity can be traced to the physicists' interest in discussions of $E = mc^2$ as one of the most testable derivatives of the fundamental, but not readily observable, changes in world view that Einstein's theory presented. Thus most of the popularizations of the 1920's devoted several pages to discussing this equation.[4]

These popularizations occasionally pointed out a curious extrapolation that could be made by a simple arithmetic calculation using the equation. What is the *total* energy in an ordinary ball? $E = mc^2$ can be used to find that total energy, by starting with a measurement of the mass of the ball.

The units to use with the equation are standard ones in physics. The mass of the ball is measured in kilograms, and a kilogram is about two pounds. A ball like a baseball has a mass of about one-tenth of a kilogram. The unit for energy in this system of measurement is called the "joule." If you toss a baseball you'll have given it about one joule of energy, in the form of motion. A 100-watt light bulb uses 100 joules per second. An automobile engine running hard needs about 50,000 joules of energy, in chemical form, every second. The total energy E, represented by the mass m of the ball, is the number to be found using the equation. The symbol c^2 is a constant number, and a huge number. In the standard system of units, that number is about 100,000,000,000,000,000 meters^2/second^2.

For a 0.1 kilogram ball, the equation becomes: E = 0.1 kilogram × 100,000,000,000,000,000 meters^2/second^2. The unit "kilogram × meters^2/second^2" is identical to one "joule." So the answer is that the energy of the ball sitting on the table in front of you is

$$E = 10,000,000,000,000,000 \text{ joules}$$

This total amount of energy in an ordinary baseball, or any other tenth kilogram of matter, is gigantic. It is enough energy to operate an automobile continuously for 4000 years. Or, if released in a split second, enough explosive energy to destroy the center of a city.

This theoretical calculation, a sidelight in early popularizations of relativity, sounds sensational. But did that calculation actually suggest that automobiles or bombs might be powered from this energy? Einstein too could calculate how much energy there was in a ball, but he didn't get excited about producing power. Because despite measures of automobiles operated or bombs exploded, almost none of the energy calculated here is in the form of mechanical or chemical energy. It is almost all locked up in the nuclei of the atoms of the ball, since the dense nucleus is where almost all the mass of any atom is to be found. There is not even a hint in the theory of relativity that it is possible, let alone practical, to release large amounts of this energy into motion, heat, or chemical forms of energy. The popularizations included the above calculation as an amusing curiosity, not as a suggestion of anything practical. Well into the 1930's Einstein himself believed that there was no way to actually convert the energy bound up in the nucleus into other forms on any significant scale.

How would one start converting the nuclear energy into heat or other useful forms of energy? What should one start with – a supply of baseballs, dynamite, or uranium? $E = mc^2$ gives no hint at all. Again, the equation does not even say it is possible to make such conversions. If such conversions were somehow produced, the equation would be useful only in hindsight, to tell how much energy was converted from a measurement of the mass lost.

One can manipulate a ball to change its energy by throwing the ball. A light toss adds only one joule to the ball's 10,000,000,000,000,000, joules of energy. Thus the mass of an ordinary ball, at rest or in motion, appears constant. A bigger energy change can be realized by burning the ball, releasing chemical energy in the form of heat. The ball can lose thousands of joules of energy that way. But that amount still represents only a tiny part of the total energy – of the order of one trillionth of one percent. Even for a ball made of a chemical that when burned releases violent amounts of energy, say a ball made of dynamite, the change in total mass-energy would remain extremely small.

Thus, while $E = mc^2$ can be correct and can function in the normal energy processes involved with our daily activities, the effects of the equation may be totally neglected. There is no suggestion in the equation itself or in any part of Einstein's work that there could be any practical applications. In 1934, twenty-seven years after he had written $E = mc^2$, Einstein was asked by a reporter if atomic energy might become a practical power source, and Einstein's reply was summarized on the front page (Fig. 4): "Atom Energy Hope is Spiked by Einstein."[5]

Thus the association of Einstein, $E = mc^2$, and the atomic bomb, enshrined together as a modern myth on the cover of *Time*, was not a direct or rapid sequence at all. The myth is a convenient expression of contemporary mistrust of science and technology, and it contains an essential truth, that basic science can produce revolutions in the broader society. A more accurate history of atomic weapons, however, offers some very different insights into the relation between science and society.

▶   *How the bomb began*

Physicists had been receptive throughout the final decade of the 19th century to the idea of a new source of energy. The stimulus was the celebrated debate between the physicists, led by Lord Kelvin (William Thomson), and the biological and geological communi-

Fig. 4. Detail from the first page of the second news section, *Pittsburgh Post-Gazette*, 29 December 1934. Reproduced by permission, courtesy of the American Institute of Physics, Niels Bohr Library.

ties.[6] Biologists and geologists were enthusiastically forming theories of evolution – theories which required billions of years of time. Kelvin did a calculation on the age of the sun, based on Helmholtz' model of the sun as a gravitationally contracting ball of gas. Kelvin concluded that the sun could have been exhausting its store of gravitational energy for no more than 100 million years, and probably for much less. Although Kelvin stuck to his estimates, astronomers and other physicists, along with the deeply distressed biologists and geologists, began to look keenly for other sources of the sun's energy so that the notion of a vastly older universe could be accepted in physics as well as in evolutionary theory. A typical speculation at the end of the century was that of T. C. Chamberlain:

> Is present knowledge relative to the behavior of matter under such extraordinary conditions as obtain in the interior of the sun sufficiently exhaustive to warrant the assertion that no unrecognized sources of heat reside there? What the internal constitution of the atoms may be is yet an open question. It is not improbable that they are complex organizations and the seats of enormous energies. Certainly, no careful chemist would affirm either that atoms are really elementary or that there may not be locked up in them energies of the first order of magnitude. [1899][7]

A discovery made in France was already confirming Chamberlain's speculation. In 1896 the French physicist A. H. Becquerel had shown that uranium ore gave off mysterious, invisible rays which could fog photographic film even in light-tight wrappers. Among the students who began dissertation projects with "radioactive" materials was Marie Curie who, with her husband Pierre, isolated several radioactive minerals in far more concentrated amounts than occur naturally. The availability of greater amounts of ray-producing materials permitted more careful measurements and experiments.

A jar of concentrated radioactive mineral slowly released enough energy in the form of heat to keep itself warm, above room temperature, for years. Marie Curie's measurements established by 1903 that the total store of energy that would eventually be released by the atoms of a radioactive rock was enormous. It was vastly more than could be stored by chemical means, or by any previously known form of energy.

Several decades were required before the precise mechanism (now called nuclear *fusion*) of energy production inside the sun could be described, yet the promised new source of energy had been found in these radioactive atoms. Darwin and the geologists were rescued from Kelvin's short-lived sun.

Among those who worked intensely with the Curie's new materials, including radium, was Ernest Rutherford, who was then in Canada. Rutherford performed careful experiments to determine the nature of the radioactive emissions, and found that some were bits of atoms, which would mean that an individual radioactive atom somehow blew itself apart releasing energy in the form of fast-moving particles. In 1903 Rutherford considered a "playful suggestion that, could a proper detonator be found, it was just conceivable that a wave of atomic disintegration might be started through matter, which would indeed make this old world vanish in smoke."[8] By 1905, Rutherford and his collaborator Frederick Soddy had created a modern theory of radioactivity.

Thus contemplation of a new energy source from radioactive atoms had begun well before Einstein's Special Theory of Relativity was published in 1905, and before $E = mc^2$. That contemplation had been stimulated a generation before Einstein by the debate over the evolutionary theories of geologists and biologists, and reached fruition, independently of Einstein, in the experimental results of the Curies, Rutherford, and Soddy.

Even the popular excitement about atomic energy had preceded the 1905 theory of relativity. The *Brooklyn Daily Eagle* devoted a Sunday page in 1903 to the story: "Radium, the Mysterious Metal, the Greatest Discovery Ever Made."[9] This popular interest was largely the result of the efforts of one man, Rutherford's collaborator Frederick Soddy.[10]

The parallels between Soddy and Einstein are noteworthy. They were born within two years of each other; received their Nobel Prizes for the same year, 1921; and died within two years of each other. But Einstein's work in 1905 didn't reach the public's attention until 1919; whereas Soddy, about the time he did his major work with Rutherford on the theory of radium and radioactivity, immediately began successful popularizing. In 1903, Soddy publicized radioactivity as a potential major source of energy.

Soddy's first public pronouncements were in 1903 and 1904 – again, more than a year before $E = mc^2$ appeared in any form:

It is possible that all heavy matter possesses latent and bound up in the structure of the atom a similar quantity of energy to that possessed by radium. If it could be tapped and controlled, what an agent it would be in shaping the world's destiny. The man who put his hand on the lever by which a parsimonious nature regulates so jealously the output of this store of energy would possess a weapon with which he could destroy the earth if he chose.[11]

And on the positive side:

The nation which can transmute matter could transform a desert continent, thaw the frozen poles, and make the whole world one smiling garden of Eden.[12]

Soddy's estimates of the energy in the atom were based on measurements by the Curies, whose work had won them the Nobel Prize, with Becquerel, in 1903. Thus the word was out – the anticipated new source of tremendous energy resided in atoms – well before Albert Einstein published anything about relativity or $E = mc^2$. The equation provided only a footnote at the time. The spur to consideration of atomic energy, for scientists and for the public, had been the experimental work with radioactivity.

Soddy's primary work of popularization, *The Interpretation of Radium*, was published in March of 1909. It was written for a lay audience, and contains no mathematics. Neither Einstein nor $E = mc^2$ is mentioned, and there is no reason to suppose that Soddy was aware of either at the time. Many dramatic passages in Soddy's book suggested significant potential consequences for society when atomic energy became a practical source:

This bottle contains about one pound of uranium oxide, and therefore about fourteen ounces of uranium. Its value is about £1. Is it not wonderful to reflect that in this little bottle there lies asleep and waiting to be evolved the energy of about nine hundred tons of coal? The energy in a ton of uranium would be sufficient to light London for a year. The store of energy in uranium would be worth a thousand times as much as the uranium itself, if only it were under our control and could be harnessed to do the world's work in the same way as the stored energy in coal has been harnessed and controlled.[13]

The Interpretation of Radium went through four editions: March 1909, November 1909, October 1912, and August 1920. In expanded and revised versions, it has remained in print for most of the past seventy years.[14]

▶ *Early uses of atomic energy in fiction*

Radioactivity, and particularly the popularizations by Soddy, quickly inspired fictional accounts of the social, political, and military impact of the eventual release of atomic energy on a vast scale. Hollis Godfrey's 1908 novel, *The Man Who Ended War*,[15] introduced what was to become a common theme. A lone scientist finds a way to speed up radioactivity, releasing enormous power at will. He blackmails the world into obeying his orders, in this case total disarmament. Even in this early piece of popular fiction, despite its awkward writing, racial stereotypes, and poorly understood physics, the potential side-effects of atomic energy are seriously explored. The balance of political power in the world can be upset by science, and a few individuals or terrorists with an atomic monopoly can wield great influence.

Garrett P. Serviss' romantic adventure, *A Columbus of Space*,[16] first published in serial form in early 1909, illustrates some lighter aspects of the potential uses of atomic energy. Serviss's hero studies the radioactivity physics of the day, and in his private laboratory constructs an atomic powered spaceship. A wild visit to Venus commences by page 11, and no further investigation of the implications of atomic energy occurs.

The tradition in fiction of deadly serious examinations of the potential of atomic energy began with H. G. Wells' science fiction novel, *The World Set Free*.[17] That novel, published in 1914, is dedicated "To Frederick Soddy's 'Interpretation of Radium.' This Story, which owes long passages to the Eleventh Chapter of that book, Inscribes and Acknowledges itself." Wells' future history describes a world transformed in the 1930's by the development of inexpensive atomic power, and then partially destroyed by atomic warfare. An early chapter, clearly based on Soddy's book, presents a lecture on radioactivity including an accurate description of the process and its potential as an energy source. One passage was lifted

almost word-for-word from the passage quoted above from Soddy's
The Interpretation of Radium:

> This little box contains about a pint of uranium oxide; that is
> to say about fourteen ounces of the element uranium. It is
> worth about a pound. And in this bottle, ladies and
> gentlemen, in the atoms in this bottle there slumbers at least
> as much energy as we could get by burning a hundred and
> sixty tons of coal. If at a word, in one instant, I could suddenly
> release that energy here and now it would blow us and
> everything about us to fragments; if I could turn it into the
> machinery that lights this city, it could keep Edinburgh
> brightly lit for a week. But at present no man knows, no man
> has an inkling of how this little lump of stuff can be made to
> hasten the release of its store.[18]

Wells' popular success inspired imitators almost immediately.
Arthur Train and Robert Williams Wood published *The Man Who
Rocked the Earth*[19] just one year later, in 1915. They cite both Wells
and Soddy, but their plot is nearly identical to Godfrey's: a lone
inventor forces the world to disarm.

These fictional accounts of atomic energy, like the popular work by
Soddy that inspired them, never mentioned either Einstein or his
equation. Einstein became widely known only after he achieved
sudden fame in 1919. It was then that writers, looking for ways to
explain Einstein's work to the public, introduced $E = mc^2$ and
connected it to the sensational speculation about atomic energy that
had been appearing for nearly two decades. In keeping with
Einstein's own views, the early popularizations were much more
conservative than Soddy and his followers had been. J. H. Thirring,
in his 1921 *The Ideas of Einstein's Theory*, applies $E = mc^2$ to a ball and
notes the enormous energy in any small amount of matter:

> This figure is stupendous, and it takes one's breath away to
> think of what might happen in a town, if the dormant energy
> of a single brick were to be set free, say in the form of an
> explosion. It would be sufficient to raze a city with millions of
> inhabitants to the ground. This, however, will never happen,
> because as we know from radio-active phenomena, these
> enormous quantities of energy contained in the nuclei of
> atoms are only liberated with extreme slowness, and are
> entirely uninfluenced by human agencies.[20]

Contemporary myth presents $E = mc^2$ as the basis for the belief in atomic energy. In fact, the opposite relationship was true. Both in Einstein's original 1905-07 work and in the popularizations of the 1920's, the prior discovery of atomic energy in the form of radioactivity provided the only experimental basis for believing that $E = mc^2$ might be valid.

After 1920 Einstein and his equation became an additional citation, along with Rutherford, Soddy, and the Curies, in the continuing assessment in fiction of the promises and dangers of atomic energy. Karel Čapek's *Krakatit* [21] (discussed in Chapter 4) in 1924 explored the dilemma of an inventor who has found the means to release atomic energy, and wishes to withhold it from a world that would surely abuse the power. Einstein is cited as an expression of the strange new world view, and not as a directly relevant figure in the moral issue. Čapek's black comedy, *The Absolute at Large*,[22] in which vast quantities of cheap atomic energy upset society, also mentions Einstein, but only to certify that nothing in the story is inconsistent with relativity.

As the novelty of vast energy itself wore thin, writers explored the political possibilities more fully. *Wings Over Europe*,[23] a 1928 play by Robert Nichols and Maurice Browne, features the conventional scientist/inventor who finds a way to release atomic energy. In this case, however, the young man offers his process to his government, insisting only that they develop immediately a comprehensive plan for peaceful use of atomic power. His politically naive gesture is spurned, and he resorts to the same threats of destruction that were effective in Godfrey's 1908, and Train and Wood's 1915 version of this story. In this 1928 version, the young scientist is assassinated and the deadly struggle continues on a less personal level among groups of scientists who have also invented atomic weapons.

A clear and accurate discussion of the relation between Einstein's $E = mc^2$ and atomic power from radioactivity appeared in John W. Campbell's Martian invasion story of 1930. "When the Atoms Failed"[24] posits invaders with atomic weapons, based on Soddy-Wells' models. But long passages of explanation correctly argue that such weapons, while releasing enormous amounts of energy, still convert only about one tenth of one percent of the mass of the atom into explosive energy. $E = mc^2$ suggests that even greater energy could be derived if *all* the mass were somehow converted to energy. An inventor succeeds in developing a total-mass-annihilation weapon, thus saving the planet Earth from the Martians. Campbell's story is noteworthy in its emphasis on the rapid pace of an atomic war. The entire great battle is fought in a single day, and a technological advantage on one side's nuclear arsenal is the deciding factor.

From the first experiments with radioactivity, some scientists and science popularizers like Soddy, and science fiction authors like Wells, considered the potential effects of the vast release of atomic energy that might someday become practical. Most scientists remained highly skeptical. The energies involved in radioactive emissions were tiny – minute fractions of a joule. But these tiny amounts were *per atom*. If, and only if, gigantic numbers of emitting radioactive atoms could be assembled or created would the energy production become big enough for practical uses. Most natural minerals emitted their energies over periods of thousands or millions of years, so only by an enormous speeding up of the process could useful rates of energy release be produced. There seemed no way to achieve this speeding up, however, and not just in the opinion of Einstein. Even Ernest Rutherford dismissed the possibility as "moonshine."[25]

▶   *The bomb becomes practical*

By the late 1920's, researchers in various parts of the world had developed particle accelerators, electric and magnetic devices that could make atoms radioactive artificially. Leaders in the development of these devices were J. D. Cockcroft and E. T. S. Walton, in England, and Ernest O. Lawrence in the United States. Practical energy production or weapons were still not serious concerns, however, since the particle accelerators consumed vastly more energy creating bits of artificially radioactive materials than could be obtained by capturing the energy released from the radioactivity.

The major advance in nuclear energy came with the discovery of a new particle, the neutron, by Englishman James Chadwick in 1932. Neutrons had in fact been involved in radioactivity experiments from the start, but because neutrons have no electrical charge (the name comes from this neutral property), they were not detected by the electrical devices physicists used to measure properties of radioactive rays. The importance of neutrons for the development of artificially radioactive materials was that neutrons were not electrically repelled from the nuclei of atoms (always highly charged electrically), and hence neutrons could smash into nuclei relatively easily, making the atoms radioactive. The energy involved in creating radioactivity using neutrons would be vastly less than that needed to operate the charged-particle accelerators, and a net gain in energy might be achieved. It remained to find a way to generate enormous numbers of neutrons, so that the rate of radioactive emissions could be artificially increased on a large scale.

Interplay between myth, fiction, and history is especially intriguing in the biography of the Hungarian physicist Leo Szilard. In 1932, Szilard had read H. G. Wells' *The World Set Free*.[26] Szilard found the book's descriptions of atomic power and then atomic war "exceedingly vivid and realistic."[27] He was keenly aware of the progress that had been made since Soddy's popularization of 1909, on which Wells' account of atomic energy was entirely based. In the autumn of the next year, Szilard recalls,

> I read in the newspapers a speech by Lord Rutherford. He was quoted as saying that he who talks about the liberation of atomic energy on an industrial scale is talking moonshine. This sort of set me pondering as I was walking the streets of London, and I remember that I stopped for a red light at the intersection of Southampton Row. As I was waiting for the light to change and as the light changed to green and I crossed the street, it suddenly occurred to me that if we could find an element which is split by neutrons and which would emit *two* neutrons when it absorbed *one* neutron, such an element, if assembled in sufficiently large mass, could sustain a nuclear chain reaction.[28]

Szilard's idea was to create, in a suitable mass of material, an internal source of neutrons. Neutrons would somehow be injected in small numbers – enough to make a few radioactive atoms in the mass. If those few radioactive atoms in turn split open promptly, emitting other neutrons, then the secondary neutrons could make another set of atoms radioactive. If each initial neutron caused the release of two secondary neutrons, and each secondary neutron caused the release of two tertiary neutrons, and on and on – then a *chain reaction* of rapidly growing intensity would have begun. Trillions of radioactive emissions would occur, releasing huge amounts of energy. The rate of neutron production would determine whether the release was even and controlled, as in a power plant, or whether the release was increasingly rapid and catastrophic, as in a bomb.

Szilard spent the next decade of his life vigorously promoting atomic energy. In early 1934 he sent Sir Hugo Hirst, founder of the British General Electric Company, pages from *The World Set Free*, with the comment: "Of course, all this is moonshine, but I have reason to believe that in so far as the industrial applications of the present discoveries in physics are concerned, the forecast of the writers may prove to be more accurate than the forecast of the scientists."[29]

Although neither Szilard nor anyone else realized it at the time, the splitting of atoms with a subsequent release of secondary neutrons had in fact been achieved in 1934. In Rome, Enrico Fermi and his colleagues had been bombarding uranium with neutrons, but they did not know how to interpret the results. Later, in Paris, Irène Joliot-Curie (daughter of Marie and Pierre Curie) did similar experiments, as did Otto Hahn, Fritz Strassmann, and Lise Meitner in Berlin. Only in 1938 did Lise Meitner and her nephew, Otto Frisch, both by then refugees from Hitler, figure out that these experiments had resulted in the splitting ("fissioning") of uranium atoms, with the release of energy in several forms, including secondary neutrons. Frisch told his colleague Niels Bohr in Copenhagen, and Bohr took the news to a large meeting of physicists in Washington in early 1939. By mid-1939, scientists in Germany, France, Italy, Holland, Britain, and the United States knew that production of nuclear energy in great quantity was an immediate possibility. They notified their governments, and with varying degrees of seriousness, Germany, Holland, France, and Britain began to take action.

On the eve of World War II, the potential manufacture of atomic weapons, even if only barely possible, had to be considered seriously. Szilard continued his efforts to interest leaders in undertaking a large scale investigation. Szilard also began a nearly successful attempt to get physicists to withhold publications that might help German scientists in the development of atomic processes. Szilard's first efforts to interest either the British or American governments were similar but even less productive than Szilard's fictional counterpart in *Wings over Europe*. The young scientist in the play has easy access to his uncle, the English Prime Minister. The young Szilard was a refugee in the United States, and he had to struggle to gain entree to the halls of political power.

Despite the continuous stream of developments in radioactivity studies for forty years, Szilard had to interest officials who had no knowledge of the possibilities of atomic energy. There had been no recent public concern, no legislative discussions, no regulations, and no treaties dealing with the potential dangers of atomic weapons. The breakthroughs in atomic physics had stimulated public interest during the first years (roughly 1896-1906). Later steady progress did not produce headlines. Frederick Soddy's warnings had received much attention during that early period, but the lack of imminent danger had cooled attention to social concerns as well as to the research itself. Soddy was disillusioned by society's indifference, and outside the field of science, only the science fiction writers still appreciated the enormous potential benefits and grave risks that Soddy had forecast. Unfortunately, science fiction was still regarded by most of the public as a pulp literature for adolescents.

Leo Szilard and a handful of physicists had the task of starting the military development of atomic weapons, and then trying to control their use, as a world war was starting. Their subsequent successes and failures were all conducted under wartime secrecy. There was no opportunity for establishing international controls, exercising public scrutiny, or injecting moral constraints based on a social consensus.

►   *Myth merges with history*

Up to 1939, Albert Einstein had had no significant impact on the development of atomic energy. The seminal papers on radioactivity in the first two decades of the century and on the recognition of fission in the 1930's rarely mention Einstein.[30] $E = mc^2$ had finally been confirmed in 1932, and was a very helpful tool to determine what energy-mass exchange had taken place in atomic experiments where the masses of the atoms before and after the exchange were known.

The significant connection between Einstein and the bomb had little to do with $E = mc^2$ or even physics, but everything to do with the mythic stature of Einstein's image in 1939.

Young refugee physicists in the U.S. such as Szilard were vitally concerned that the development of atomic weapons was taking place in Hitler's Germany as well as in the allied nations. A key resource in atomic research was the element uranium, which was uniquely suitable for the work. Szilard, and another physicist, Eugene Wigner, discussed ways to restrict Germany's access to uranium. The largest stocks of the mineral came from the Belgian Congo, and it just so happened that the world's best-known scientist, Albert Einstein, had a warm friendship with Queen Mother Elizabeth of the Belgians. Perhaps Einstein's fame and his friendship could be used to alert allied leaders to the importance of that uranium.

Wigner and Szilard visited Einstein in July 1939, and their discussions led to the decision that they should notify Washington before approaching Belgium. It occurred to the younger physicists that Einstein's name might also be the entree to the Washington power structure that the less well-known younger men needed. A few weeks later, Szilard returned with another refugee, Edward Teller, and a letter to Roosevelt was prepared. That famous letter, signed by Einstein and dated August 2, warned that "the element uranium may be turned into a new and important source of energy. . . ." That letter is a serious basis for connecting Einstein with the development of the atomic bomb.[31]

Einstein was involved not because he had made contributions to the development of atomic energy or had supported that work (recall his "spiking" of the hope of atomic energy in 1934), but because his fame would facilitate reaching world leaders. The political sensitivity of Szilard and his colleagues is indicated by the concern they felt for such details as how the letter was to be delivered. Charles Lindbergh, an aviator and national hero, was the first choice for that task, but they eventually settled on economist Alexander Sachs, a personal advisor to President Roosevelt.

The physicists knew that Einstein was useful because of his image, conjuring up the full powers of science in an age when the applicability of theoretical physics to practical warfare was still questionable. Alexander Sachs describes Einstein's contribution with terms appropriate to a mythological being: "We really only needed Einstein in order to provide Szilard with a halo, as Szilard was almost unknown in the United States."[32]

Today's popular myth has taken that literal link, as well as the much less plausible connection between relativity and the bomb, as the central example in a vision of the relations between science and society. After the bomb was dropped on Japan in 1945, science itself took on a new and darker significance in the world. Albert Einstein, who had provided the image of the ultimate scientist with his theory of relativity, now became a key image of the new danger posed by science. Einstein's links to the bomb, both the vastly overstressed association of $E = mc^2$ and the authentic one of the letter of 1939, were a convenient focus for discussions of the moral responsibility or irresponsibility of science.

The shift in attitudes toward Einstein, and science in general, can be seen in both serious and popular literature. In 1930, George Bernard Shaw praised Einstein, who, in contrast to makers of empires, was a "maker of universes," with the blood of no man on his hands.[33] Rebecca West was inspired by Shaw's speech to publish an essay in 1931, "Blessed are the Pure in Heart," which closes with an image of Einstein. West clearly means no irony in her depiction of Einstein's innocence:

> One perceived that the hierarchies of the earth, in spite of his high place among them, were invisible to him. He held the stuff of life towards the light in some way so that that kind of embroidery did not show....a vision of a universe incredibly easy and roomy to live in, because nothing was suffered to exist in it but that which was real.[34]

The 1969 Time-Life documentary film, *Einstein*,[35] uses the same speech by Shaw for a terrible irony, as Shaw's tribute to Einstein is replayed after a recounting of Einstein's "connections" (especially $E = mc^2$) to the atomic bombing of Japan.

▶   *The myth in fiction and non-fiction*
In the decade preceding World War II, Einstein was the personification of science, and atomic physics was a poorly understood but central topic of science. $E = mc^2$ was a short equation and even if it was not the totality of his great theory of relativity, the public took it to be so. Since 1919, when Einstein had become famous, scientists and science fiction writers discussing atomic energy had used $E = mc^2$ to illustrate how much energy was involved with radioactivity. Soddy had long ago dropped from public view, and Szilard and the other scientists working on atomic energy were hardly known to the general public.

The mythical direct connection between Einstein, relativity, and the threat of nuclear holocaust was thus easily made immediately after World War II. The *Time* cover story of July 1, 1946, (Figure 3) contains all the elements, both factual and fictitious, of that connection. The essay begins with photographs and a description of the planned Bikini Island atomic bomb tests. Then: "*The Genius.*

Through the incomparable blast and flame that will follow, there will
be dimly discernible, to those who are interested in cause & effect in
history, the features of a shy, almost saintly, childlike little man with
the soft brown eyes, the drooping facial lines of a world-weary hound,
and hair like an aurora borealis."[36] The biographic sketch of Einstein
and his supposed connections to the bomb present the full legend.
The irony is heavy. Saintly Einstein is responsible for unleashing the
ultimate destructive force. While admitting that Einstein did not
work directly on the bomb, the article insists that the story is still
valid: "But Einstein was the father of the bomb in two important
ways: 1) it was his initiative which started U.S. bomb research; 2) it
was his equation ($E = mc^2$) which made the atomic bomb theoreti-
cally possible."[37] The article dramatically states the new concern for
the relations between intellect and society, now symbolized by
Einstein's experience: "If the atom bomb blasted the last popular
skepticism about Einstein's genius it also blasted man's complacent
pride in the power of unaided intellect. At the very moment that it
was finally mastered, matter was most elusive and most menacing."[38]

The modern myth about science and society, described at the
beginning of this chapter, has two facets which are clearly present in
the *Time* essay on Einstein and the bomb. The first facet is the
potential of science to liberate great power, represented by Einstein's
$E = mc^2$, making the bomb possible. As this section has shown, while
$E = mc^2$ became our symbol for atomic energy, it was not a cause. But
fundamental science did lead to the release of that energy and then to
building of the bomb, so the myth has a general validity despite the
inaccuracy of the central example. The juxtaposition of the three
images – Einstein, the equation, and the bomb – is now a widely
recognized icon, repeated over and over again in post World War II
literature, both fiction and prose. A full page picture of Einstein in
The Unicorn Book of 1952 is captioned: "*Father of Atomic Era*: Albert
Einstein's theory of relativity made nuclear fission possible...."[39]
Nearly three decades later, the same inaccurate image still appears
widely. In a 1981 book of popular facts, the paragraph on "Relativity" informs readers: "One of the many aspects of the theory is that
mass and energy are equivalent, a basic concept leading to the
development of nuclear fission and the atomic bomb. The bomb
demonstrated most impressively the divisibility of the atom....
However, it may some day render the concepts of psychology,
evolution, and the heliocentric system irrelevant, since there may be

no one left to psychoanalyze, nothing to evolve, and nothing left to revolve around the sun but radioactive dust."[40]

The second aspect of the general myth is that powers are released by science so suddenly that society has no chance to consider the potential dangers. Einstein's image is of an old, tired man, full of great sorrow for the harm he has inadvertently released. As we have seen, in the period of fifty years between the first research on radioactivity (and coincidentally $E = mc^2$) and the realization of the bomb, the potential dangers and benefits of atomic energy were widely discussed, not only by scientists, but by popularizers and fiction authors. In this case, the central example is not only inconsistent with the myth, but is a significant counter-example.

Beginning with Hollis Godfrey and H. G. Wells, authors of fiction had considered the general classes of side effects: economic, political, military, and social. In the early 1940's, the discussions in science fiction ran ahead of the developments in science in technical details as well. Nuclear plant accidents were vividly discussed in terms of the pressure on operators in Robert Heinlein's "Blowups Happen," published in 1940.[41] The first actual reactor was operated secretly in Chicago in 1942. Theodore Sturgeon's 1941 "Artman Process,"[42] considered a struggle between major powers to steal a process for separating fissionable uranium 235 from uranium 238, the dominant form in natural ore. The development of such a process was the major industrial activity of the Manhattan Project, officially begun in 1942 amidst great concern that the Germans were working on the process as well.[43] Heinlein's "Solution Unsatisfactory," published under a pseudonym in 1941,[44] contemplated the need for a world-wide dictatorship which would control nuclear weapons as the only means of preventing total destruction. Lester Del Rey's "Nerves" examined a fictional runaway nuclear reactor in 1942.[45] In 1944, Cleve Cartmill's "Deadline" described the mechanism of an atomic bomb with such detail that the Federal Bureau of Investigation looked into the possibility of a security leak.[46] That same year, Clifford Simak's "Lobby" examined the politics and economic struggles that might accompany the introduction of commercial nuclear power.[47] Immediately after the war, science fiction stories abounded about the aftermath of a world-wide nuclear holocaust. Chan Davis' 1946 "The Nightmare" considered a new threat: nuclear terrorists, backed by a foreign power, who assemble a bomb in New York City with the intention of blackmail.[48]

These stories were not startling as science prognostication; they used the most recent discoveries and inventions and applied straight-forward extrapolation to predict the possible social consequences. Some of the stories are poorly written, with elements of pulp adventure competing with serious social commentary. Nevertheless, the attention paid by science fiction to the potentials of atomic energy, from 1908 until the actual utilization of atomic energy in 1945, is a convincing demonstration that the implications of *this* power released by science could have been thoroughly and publicly considered. The aspect of the general myth about science and society, that change occurs too rapidly for careful study, is contradicted by the decades of serious and popular examination of atomic energy.

▶   *The myth in later fiction and drama*
In post-war fiction, we can find Einstein used to illustrate achievement turning to tragedy. Pierre Boulle's short story, "$E = mc^2$," presents an accurate history of the development of the bomb, but with a twist – all the scientists involved (Fermi, Bohr, Einstein, and others), some with their names slightly altered, think of using nuclear physics as a gesture of peace.[49] They propose to convert energy into mass, and show the world that "$E = mc^2$ is the very symbol of love."[50] Their great effort culminates with the conversion of cosmic ray energy into uranium blossoms over Hiroshima. Something goes wrong, tragedy results, and Einstein groans "*I was the one who pressed the button*," while the Fermi-character laments, "God knows I didn't want this!"[51] The scientists are responsible, and their failure was lack of control over the ultimate results of their work.

In his *Galileo*, Bertolt Brecht, who called himself the "Einstein of the new theatrical form,"[52] also raises the questions of scientists' responsibility for their ideas. Although the play explores the Renaissance struggle between the Church and Galileo, the issues deliberately pertain to modern scientists and the decisions that must be made about the use of their inventions and discoveries.

Actually the play has three versions. The original, written in 1938, showed a heroic Galileo fighting for progress against the Inquisition. A second version was written in 1944-45 with Charles Laughton, who helped turn the stage Galileo into a coward and self-centered opportunist. In the preface to this American edition, Brecht wrote: "The atomic age made its debut in Hiroshima while we were in the midst of our work. Overnight the biography of the founder of modern physics read differently. The infernal effect of the great bomb placed the conflict of Galileo with the authorities of his age in a new and sharper light."[53] The numerous changes in this second version, particularly in the recantation scene, connect Galileo with contemporary physicists whom Brecht held largely responsible for the nuclear nightmare. A third version of the play makes Galileo an ambiguous composite of the first two, a more complex and interesting figure, even though the first version was closer to historical fact.

Another modern dramatist to explore the consequences of physics was Friedrich Dürrenmatt whose play *The Physicists* was the most frequently performed play of the 1962-63 season in the German-speaking world.[54] Behind the masks of pretended insanity, characters named Newton, Einstein, and Möbius contend for control of scientific secrets related to nuclear power. But they learn that man is not prepared to control his own scientific discoveries nor can he keep his discoveries hidden from the socio-political forces which seek them. The play is based partly on Leo Szilard's attempt to hide his knowledge of the chain reaction process and of his attempts during the war to prevent the dropping of the atomic bomb on people.

Before writing *The Physicists*, Dürrenmatt reviewed Robert Jungk's *Brighter than a Thousand Suns*[55] – a history of the development of the atomic bomb which includes accounts of Szilard's trying to bury his scheme for chain reactions in the British Admiralty and accounts of Szilard's and other physicists' efforts after the fall of Germany to stop the bomb from being used.

In the play, the three physicists, although sane, are inmates of an insane asylum. Hoping to keep their dangerous secrets safe from society, they try to be "physicists but innocent,"[56] as if that were a contradiction in terms. And it is, in a sense. The insane director of the asylum has already stolen the secrets the physicists thought they were safeguarding, and she will bring about a final apocalypse as she exploits the science for her own ends. Möbius envisions "somewhere round a small, yellow, nameless star there circles, pointlessly, everlastingly, the radioactive earth."[57] The play ends with "Einstein," a prisoner in a madhouse (human society), hopelessly playing his violin. In spite of the best efforts and great personal sacrifices, the scientists have not been able to prevent their theories and discoveries from being abused by power-hungry businessmen and politicians.

A similar tragic conclusion, with Einstein himself making an appearance, is presented in the 1957 *X Minus One* radio play, "Target One."[58] The play opens in 1990 after a third world war has destroyed most of civilization. The "President of the World Council" offers a scientist a fantastic opportunity: the chance to use a time machine to prevent the war which has already happened. The President asks the Scientist:

> P: What would you say was the most
>   important scientific discovery of the 20th century?
>
> S: I would suppose Einstein's
>   equation, $E = mc^2$.
>
> P: Exactly. This formula unlocked the
>   atom. Without it there would have been no atomic energy, and consequently, no bomb. No war.
>
> S: The formula did not cause the war
>   Mr. President. Nor did the man who discovered it. He was one of the finest human beings who ever lived – a man utterly dedicated to the pursuit of peace and harmony.
>
> P: Yes, exactly, that is what will
>   make your task more abhorrent, doctor.
>
> S: What is this task Mr. President?
>
> P: Quite simply, it is the murder of
>   Albert Einstein.

The time machine works, and Einstein is murdered before he has a chance to invent the dreaded equation. Yet the third world war is not prevented, only delayed. Somebody else came up with $E = mc^2$, and all else proceeded to the same conclusion. Although this play script repeats the drastic exaggeration of the role $E = mc^2$ played, its conclusion, as in the historically more perceptive story by Boulle, is the same. Atomic destruction seems inescapable, for knowledge itself contains the seeds of destruction.

▶   *Inevitability*

Inevitability of scientific and technological developments was not a notion devised just to separate Einstein from the ultimate responsibility for the bomb. Science has often been perceived as a strict progression in uncovering facts about the universe, a process in which the personalities of the individuals doing the uncovering are of merely temporary importance. While many historians of science (Kuhn, Bronowski, and Holton, for example)[59] disagree with that proposition, the release from individual responsibility has always been a potent attraction. H. G. Wells' 1914 science fiction work about nuclear war includes an inevitability argument which is still being used by characters in the fiction by Boulle, Brecht, Dürrenmatt, and the author of "Target One," after half a century, and after the actual fact of atomic warfare. In Wells' version the creator of induced radioactivity in 1933 is a physicist, Holsten, (the experiment was in fact done by Irène and Frédéric Joliot-Curie in that same year) and the possibility of any other outcome from his experiments is dismissed in this manner:

> There is a kind of inevitable logic now in the progress of research. For a hundred years and more thought and science have been going their own way regardless of the common events of life. You see – *they have got loose*. If there had been no Holsten there would have been some similar man. If atomic energy had not come in one year, it would have come in another.[60] [Emphasis Wells']

In Dürrenmatt's play the psychiatrist-villain who is about to destroy the world invokes inevitability to explain why the men who tried to keep dangerous knowledge secret had to fail: "For what was revealed to him was no secret. Because it could be thought. Everything that can be thought is thought at some time or another. Now or in the future."[61] The characters, villains and would-be heroes, are deeply flawed so that the misuse of information is the result of a moral weakness and madness, not merely intellectual inevitability. Nevertheless, the end is destruction once again.

The physicists who developed the atomic bomb echo their predecessors and followers in fiction. Journalist Langston Lamont interviewed dozens of the physicists involved in the Manhattan Project and the first atomic explosion ("Trinity") and reported:

> Several strains run through the conversations of the scientists as they reminisce on Trinity and the decision to drop the bomb. One is the *inevitability* of the whole affair, from the first test to Hiroshima and Nagasaki. The word recurs over and over in the explanation that science was bound sooner or

later to unlock the atom and that it was fortunate the first honors fell to America and not to Germany or Russia. It was *inevitable* that the atomic bomb would be tested and that it would be used in battle.[62] [Emphasis Lamont's]

Einstein's supposed origination of the bomb, and his subsequent release from personal responsibility afforded by the inevitability argument, was not then a unique sequence of the Einstein legend. But Einstein makes a convenient embodiment of genius that may turn to less than noble ends. The Einstein myth (even with its historical inaccuracies) serves as an invaluable story to focus the continuing struggle to understand the relations between science and society. One wonders if the struggle would proceed more fruitfully if a more accurate myth were the most cited example of those relations.

▶   *Einstein as a personification of intellect*
In addition to its presumed role in atomic energy, the public image of Albert Einstein has come to represent intelligence in general, and the scientific mind in particular. That image is consistent with a national stereotype of scientists as distinctly odd people. In 1956 a national poll of high school students found that between 25 and 30 per cent reported extreme views such as that good scientists must be geniuses, cannot expect to raise a normal family, are "more than a little bit odd," and do not have time to enjoy life.[63]

Portraits of Einstein from mass circulation material in the 1970's present certain prominent recurring features. Einstein appears throughout as a very old man; the symbols $E = mc^2$ are used to represent his major work and nuclear power; and Einstein is used to represent supreme intellect, which is out of reach of the ordinary mind. The misleading or inaccurate nature of these features has not diminished their importance in making a powerful cultural myth out of Einstein's image.

The centennial of Einstein's birth provided numerous examples of the uses of his image. Fig. 5 is the United States' Einstein centennial stamp from 1979, and Fig. 6 is his portrait from a stamp by the People's Republic of China.[64] That stamp includes, in western letters, the formula $E = mc^2$. Most users of the stamp would not be expected to understand the English alphabet, let alone the scientific meaning of the letters. The potency of the equation does not depend on the actual physics it represents, and the physics, as discussed at the beginning of this chapter, is as equally mysterious to the western

Fig. 5. Einstein on a U.S. postage stamp. © U.S. Postal Service, reproduced by permission.

Fig. 6. Einstein on a postage stamp from the Republic of China, 1979.

public. The centennial year saw not only official government celebrations of Einstein, but a flood of attention in the popular media. *Look* magazine featured a collection of offbeat photographs of Einstein, billed on the cover in a curious juxtaposition: "The Unknown Elvis/The Unseen Einstein."[65] The scientist and the pop singer rated equal size type, and one might consider which one has been honored by the equal treatment. Fig. 7 is a more intentionally humorous presentation. It's a fashion catalog and has nothing to do with physics, but nevertheless there's the elderly face of Einstein. On another page of the catalog, Fig. 8, the face is surrounded by $E = mc^2$, along with text about choosing apparel components wisely.

An advertiser doesn't need to explain the use of Einstein's picture, because when we see him the association with intelligence or smart

Fig. 7. Cover of a Pennaco Hosiery catalog, 1979. Reproduced by permission, courtesy of Barbara Guzy, Pennaco Hosiery Division, International Playtex, Inc.

thinking is immediate. To picture genius, we imagine Einstein. The same theme is evident in Fig. 9: "Instinct Says Beer/Reason Says Carlsberg."[66] Einstein is the embodiment of reason.

Business computer users, targets of the amusing advertisement in Fig. 10, are not expected to appreciate subtleties of science – they will not care that theories are invented, not discovered, and that large

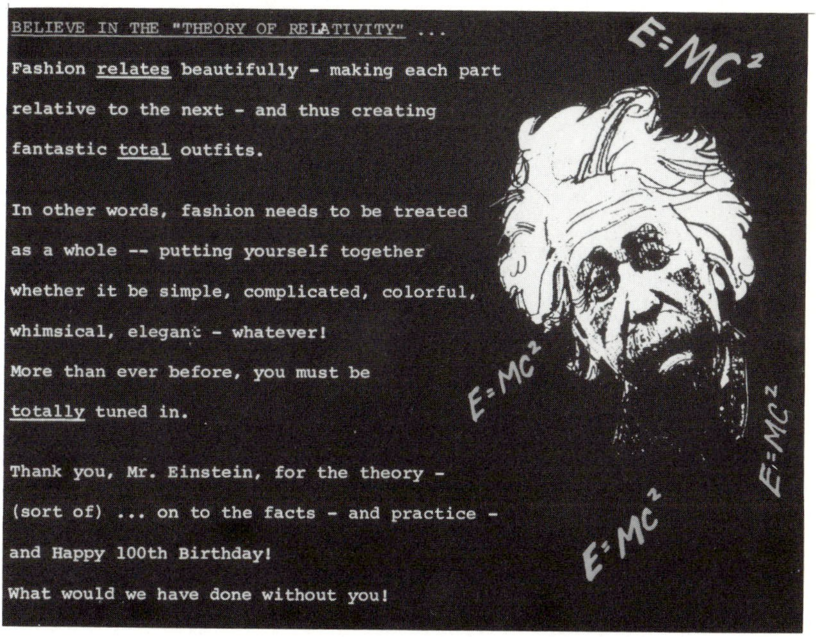

BELIEVE IN THE "THEORY OF RELATIVITY" ...
Fashion relates beautifully - making each part relative to the next - and thus creating fantastic total outfits.
In other words, fashion needs to be treated as a whole -- putting yourself together whether it be simple, complicated, colorful, whimsical, elegant - whatever!
More than ever before, you must be totally tuned in.
Thank you, Mr. Einstein, for the theory - (sort of) ... on to the facts - and practice - and Happy 100th Birthday!
What would we have done without you!
E=MC²
E=MC²
E=MC²
E=MC²

Fig. 8. Detail from page 1 of the Pennaco Hosiery catalog.

l'istinto dice birra.
la ragione dice Carlsberg.

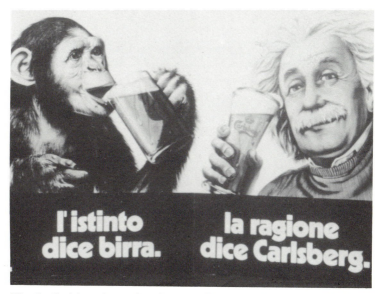

Fig. 9. Carlsberg beer advertisement. Reproduced courtesy of J. Goodstein, California Institute of Technology.

Fig. 10. Data General advertisement, © 1975 Data General Corporation, and reproduced courtesy of Data General.

scale number manipulation (the strength of these computers) was not useful in developing relativity. For this image to work, the reader needs to understand only that Einstein means thinking power, and that this company's wares have that power too. Several other features of this advertisement clearly illustrate recurrent aspects of the Einstein myth. As a very old man, he is in the act of discovering relativity, which culminates in the equation $E = mc^2$. And he does not have time (especially in old age) to explain the theory.

"Normal people do not have the intelligence necessary to understand Einstein." Some version of that feature of the myth is perhaps most useful for promotors claiming comparative simplicity for their arguments. The statement that only twelve people in the world understand relativity has reappeared in print continually for the past fifty years, although more recent popularizations have begun to question its validity.[67] The continuing universality of that myth is illustrated by a tribute to Einstein in a popular Indian weekly in 1978: "'Everyone knows that Einstein did something astounding, but very few people know exactly what it was that he did.' Bertrand Russell opened his book, *The ABC of Relativity*, 53 years ago with these words. After nearly 75 years since the publication of Einstein's Special Theory of Relativity in 1905, Russell's observation still holds good. When I was a schoolboy, it used to be said that those who really understood the Theory of Relativity were probably fewer than a dozen."[68]

Keeping this impression of genius beyond normal understanding in mind, the humor of the following advertisement becomes clear. Fig. 11, Einstein holding an electric bill before his eyes, clearly states that one does not need super-human intellect to decipher the mysteries behind utility bills. Indeed, too much genius is a hindrance, as Fig. 12 suggests. The computer company's staff has a reputation of being made up of Einsteins, but that is a reputation the company wishes to take pains to correct. Why is it *bad* to have Einsteins on your staff? Because normal folks can't understand them. In particular, we are assured, purchasers of the company's computers will not need to strain to understand their new tools.

That genius like Einstein's is a mixed blessing is explicit in the use of his portrait to represent a questioned goal in Fig. 13. This illustration is from a brochure for a documentary filmstrip on genetic manipulation. The brochure asks what moral dilemmas we would face if cloning of humans became possible. Multiple images of

You don't have to be a nuclear physicist to understand an electric bill.

Fig. 11.   Detail from an advertisement. © 1977, Allied Chemical Corporation, and reproduced by permission from Allied Corporation.

Einstein first suggest that if we were to reproduce valuable people, Einstein would leap to mind, but, on reflection, do we want more Einsteins? Einstein is our symbol of mankind's intellectual achievement, yet we fear intellect that is beyond most of us. After all, the myth goes, that intellect produced $E = mc^2$, cause of nuclear weapons and possible global holocaust.

Einstein's aged face has come to represent the ambiguity of our attitude toward intellect: respected yet feared, a force that is mysterious and perhaps not under control. Fig. 14 is the real Einstein at the time he was completing the Special Theory of Relativity. He's not a frail old man at all, but a rather handsome young fellow, twenty-six years old. This young man published four papers in 1905, stating the Special Theory of Relativity, the quantum theory of light, and two other important essays on atomic physics. The Einstein who "discovers" relativity in popular image is at least forty years older than the Einstein who invented relativity in fact. The association of wisdom with age is a longstanding notion, however, and so if Einstein is to represent intellectual wisdom for our culture, he had better be much older than a tender twenty-six years. When Einstein completed his most famous works, he was thirty-seven, and the next year he wrote a popularization entitled *Relativity*.[69] Today, relativity is taught and understood qualitatively in thousands of introductory physics classes.

We want
to get rid of a good reputation

It has become well known that Norsk Data has delivered a number of NORD Computers to Universities, laboratories and research institutes. So well known has this become that many think this is all we do. But it is not.

With our powerful NORD-10/S computer, SINTRAN III operating system and full range of peripheral equipment, we can offer flexible and economic configurations for most data-processing activities. We have especially concentrated on making systems which are easy to apply. In this way smaller firms can have their own data centre and develop their own application programs without always needing expensive assistance from outside consultants.

Give us a call and find out how we can co-operate.

Norsk Data

Lørenveien 57, Postboks 163,
Økern, Oslo 5.
Tel. (02) 21 73 71/22 80 90.

Sweden: Norsk Data AB, Box 2031, 19402 Upplands Vasby. Tel.: 0760-86 050.
Denmark: Norsk Data ApS, Øverødvej 5, 2840 Holte. Tel.: (02) 42 50 55.
France: Norsk Data s.a.r.l., 64 Rue de Meyrin, 01210 Ferney-Voltaire. Tel.: 50 41 65 41.
England: Richard Norton (NORD) Ltd. NORD House, 17 Balfe Street, London N1. Tel.: 01-2785501, Telex: 299 537.

Norsk Data A.S develops, produces and markets NORD Computer Systems. The NORD systems have found extensive use in research and education process control, data processing and networks; the application areas are continuously increasing.

Fig. 12. Norsk Data advertisement of 1978, reprinted by permission from Norsk Data S.A.

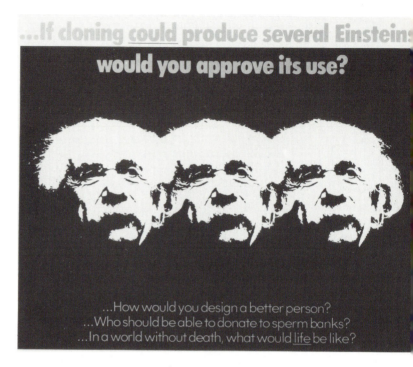

...If cloning could produce several Einstein
would you approve its use?

...How would you design a better person?
...Who should be able to donate to sperm banks?
...In a world without death, what would life be like?

Fig. 13 Harper and Row filmstrip advertisement, 1977.

▶   *Image at the fringe of culture*

Awareness of Albert Einstein spread from the remote society of theoretical science out to advertisements and fiction. The invocation of Einstein at the outermost fringes of culture as well establishes a strong case for Einstein having become a universal cultural icon. Delusions and fantasies having Einstein as a central image tell us almost nothing about the scientist, if much about the fantasizer, but these images do show how thoroughly society can extract the mythic content from a symbol.

Anti- or post-Einstein articles and books appear regularly, usually the work of amateur scientists who deal with ideas of relativity but not with its more difficult mathematical aspects. These works are often based on demonstrations of flaws and limitations in the analogies used in conventional popularizations; analogies never intended by their inventors to be more than rough approximations. Nevertheless, failures in the analogies are taken to be failures of Einstein's theories. The effort and commitment given to overthrowing "establishment"

Fig. 14. Einstein in 1905. Reproduced by permission of the Hebrew University of Jerusalem, courtesy of the American Institute of Physics, Niels Bohr Library.

relativity can be substantial. In one instance, a full-page advertisement in a Berkeley, California newspaper, headlined "SCIENCE
STUDENTS,   FACULTY   MEMBERS/$1,000   REWARD,"
offered payment for an experiment disproving one particular basis of
Einstein's work.[70]

While scientists do not take these attacks seriously, physicists
continue their practice of skepticism toward even the most securely
established concepts. Experimental tests of relativity and alternative
theories – tests which so far entirely support Einstein's version – are
nevertheless seen by anti-Einstein enthusiasts as encouraging their
activity.

Two more extreme cases, involving paranoid delusions, will
complete this survey of the Einstein myth. These two examples
reduce the Einstein icon to its most potent element – the notion of an
intellect so far beyond human standards that god-like power becomes
available.

Herbert Mullen was born on April 18, 1947, in Salinas, California.
Twenty-five years later, Mullen was a psychotic, a paranoid, and a
mass murderer. Mullen attached great weight to two coincidences.
Mullen's birthday of April 18 was the date in 1906 of the great San
Francisco earthquake, and on another April 18, 1955, Albert
Einstein died. Donald T. Lunde, a psychiatrist who has studied
Mullen and testified at his trial, wrote about Mullen that:

> He also claimed that the death of Albert Einstein on April
> 18th, 1955, had somehow prevented him from being killed in
> Vietnam, but I was never able to follow through this line of
> thought. Testifying at his trial in 1973, Herb rambled about
> the "birth-position-placement-director" who has the power
> "to put people's entities in bodies that would be in a combat
> zone and then those bodies would be killed." Apparently
> Herb believed his father had made such a petition to the
> "placement director," but it was negated by Albert Einstein's
> sacrificial death on Herb's behalf.[71]

At his trial, Mullen explained why he had killed five people in the
Santa Cruz hills.

> I, Herb Mullen, born April 18, 1947, was chosen as the
> designated leader of my generation by Professor Dr. Albert
> Einstein on April 18, 1955.... His hope probably was that the
> April 18th people would use his designation and its resulting

power and social influence to guide, project, and perfect the resources of our planet and universe.... *One man consenting to be murdered protects the millions of other human beings living in the cataclysmic earthquake/tidal area.*[72]

The myth of Albert Einstein here is a general image of dreadful, inhuman power, and that aspect of Einstein's image has become a focus for Mullen's madness.

Another tragedy involved Donald Crowhurst, an engineer in Britain who owned a small company that made electronic navigation equipment.[73] His company was failing, and as with everything else in his life, it had never quite worked. So Crowhurst decided to gamble everything on becoming famous, and making his company famous, by entering a single-handed yacht race. The idea was to go around the world in a small boat without stopping. Crowhurst forgot to load many essential supplies, but took along with him things that made him feel comfortable, including electronic parts, and Einstein's popularization, *Relativity*.

Crowhurst sailed into the South Atlantic and became terrified: What if the boat capsized? And so he began to cheat. Instead of going around the world, he began sailing in circles in the South Atlantic. He began inventing very elaborate logbooks to show that he was circumnavigating the globe. Towards the end of the race he pretended he was turning around and coming back home, and he did start sailing back to England along with the leaders in the race. Crowhurst realized that when he got into port the judges would find mistakes in his false logs, so he decided to come in second, still achieving publicity, but avoiding most of the scrutiny. Then the leader dropped out. Crowhurst could not avoid reaching England in first place.

This remarkable saga is well documented. Crowhurst was making 16 mm film and tape recordings for the BBC, and left a record of a man going mad. His final comfort was through a curious reading of Einstein. From the diaries, Crowhurst's chroniclers surmise:

> It was more than comfort, it was the revelation. Einstein offered him the perfect way to cope with his nightmare! When Einstein was faced with a mathematical impasse he merely "stipulated of his own free will" that the impasse should disappear. If Einstein could make something so, merely because he *wanted* it to be so, then so could

Crowhurst. He had the free will to conjure up anything he wanted! And what he wanted was escape from this cabin and this predicament.[74]

Crowhurst's own journal includes:

The extrovert, say a politician, places his fulcrum nearer the load, for his function is to move the whole politico-economic system of his country – perhaps of the world. Both types of activity shape the course of man's history. The first shattering application of the idea that $E = mc^2$ is a good example of this – I refer to the bombing of Hiroshima.

...Einstein – a Jew – the face of God, or Christ – the Messiah? The King that brings Salvation to the Jews? Nuclear Power! Mystery of Prophecy![75]

Crowhurst committed suicide by stepping over the side of the boat. His Einstein fantasy apparently made that resolution of his dilemma easier.

▶   *Status of the mythic image*

We have examined the image of Albert Einstein as a cultural icon, and seen the range of uses of that image, from the humorous to the tragic. Einstein's image has not by itself created the vision of science in 20th-century culture, but it has been adopted by that culture to represent the intellectual enterprise.

In some way, however, the real Einstein has at least modified the image of science. In pre-20th-century debates on the moral responsibility of science, symbolized by Mary Shelley's Frankenstein, the dangers to society from the scientist were in large part due to the classic sins of man: pride, greed, selfishness. But the conception of Einstein is closer to that of a saint – he is seen to radiate cosmic humility, generosity, unselfishness. The *Time* magazine Einstein centennial cover from 1979 (Fig. 15) illustrates this aspect of Einstein's image. Gone are the mushroom cloud and warships of the 1946 cover. An Einstein of the same approximate age as shown in 1946 now has galaxies and nebulae for an environment. Thus the evil coming from science, exemplified by the transformation of gentle Einstein's $E = mc^2$ into the terror of atomic holocaust, can be seen as a failing not of the man, but of the society, or of society's abuse of knowledge.

Fig. 15. *Time* magazine cover for 19 February 1979, by Roy Anderson. ©
1979, Time Inc. All rights reserved. Reprinted by permission from *TIME*.

By limiting its image of science largely to one individual, our
culture has also reinforced some of its specific prejudices about
intellect. First, no other age, sex, or skin color seems to fit so well with
genius, so that a young, black woman must overcome an image as well
as more concrete barriers, if she is to be encouraged and recognized as

a scientist. Discrimination in education and in science is certainly not supported by Einstein's biography, but the use of Einstein as a sole image makes it easier for society to educate youth to maintain old predilections. The same sole-image prejudice attaches Einstein's foibles and failures (particularly in politics) to science and genius as a whole.

Second, since Einstein's field of science was mathematical physics, that realm of endeavor is *the* prestige field in which to look for great intellectual achievement. The arts, poetry, politics, and even other sciences are perceived with less status, in part due to Einstein's eminence. We look to physical science for solutions to societies' problems with more seriousness, and more money, than we might if our culture had happened to choose a philosopher, a poet, or a psychologist for its symbol of intellect.

And finally, since Einstein is not supposed to be understandable to normal intellects, Einstein's image supports the anti-intellectual notion that higher thought is not fit for general public consumption. C. P. Snow's two cultures have been kept apart by conditioning that begins in grammar school, when the youngster who expresses interest in science is taunted with the challenge – "Who do you think you are – an Einstein?"

This survey of the cultural image of Albert Einstein raises questions of society's response to ideas, and to dramatically gifted individuals. Certainly Einstein's person and image have been used to express and support concepts, such as "everything is relative," which in reality he would never have accepted. The myth of science unleashing powers without warning has hindered rational consideration of the role of science in society. Society's guilt over such processes as the evolution of nuclear weapons is inappropriately assuaged if a single, saintly member of society can somehow be imagined to bear a large share of the responsibility. These facets of Einstein's existence have been far more powerful in the culture at large than have his ideas and his actual biography. The uses and celebrations of those realities in works by many artists and authors examined elsewhere in this book demonstrate that fruitful inspiration can come from fact as well as from myth. But Einstein's work and words have been used less in this final chapter, dealing with the broader cultural influence of Albert Einstein's life. Image apparently has wider penetration in our culture than idea, so that society at large has accepted only the image of genius, not the genius itself, of Albert Einstein.

APPENDIX

▶ WILLIAM CARLOS WILLIAMS:

St. Francis Einstein of the Daffodils

In March's black boat
Einstein and April
have come at the time in fashion
up out of the sea
through the rippling daffodils
in the foreyard of
the dead Statue of Liberty
whose stonearms
are powerless against them
the Venusremembering wavelets
breaking into laughter –

Sweet Land of Liberty,
at last, in the end of time,
Einstein has come by force of
complicated mathematics
among the tormented fruit trees
to buy freedom
for the daffodils
till the unchained orchards
shake their tufted flowers –
Yiddishe springtime!

At the time in fashion
Einstein has come
bringing April in his head
up from the sea
in Thomas March Jefferson's
black boat bringing
freedom under the dead
Statue of Liberty
to free the daffodils in
the water which sing:
Einstein has remembered us
Savior of the daffodils!

A twig for all the dead!
shout the dark maples
in the tearing wind, shaking
pom-poms of green flowers –
April Einstein has come
to liberate us
here among
the Venusremembering daffodils
Yiddishe springtime of the mind
and a great pool of rainwater
under
the blossomy peachtrees.

April Einstein
through the blossomy waters
rebellious, laughing
under liberty's dead arm
has come among the daffodils
shouting
that flowers and men
were created
relatively equal.
Oldfashioned knowledge is
dead under the blossoming peachtrees.

Einstein, tall as a violet
in the latticearbor corner
is tall as a blossomy
peartree! The shell
of the world is split
and from under the sea
Einstein has emerged
triumphant, St. Francis
of the daffodils!

O Samos, Samos
dead and buried. Lesbia is
a black cat in the freshturned
garden. All dead.
All flesh that they have sung
is long since rotten.
Sing of it no longer.

Sing of Einstein's
Yiddishe peachtrees, sing of
sleep among the cherryblossoms.
Sing of wise newspapers
that quote the great mathematician:
A little touch of
Einstein in the night –

Side by side the young and old
trees take the sun together,
the maples, green and red
according to their kind,
yellowbells and the
vermillion quinceflower together –
The tall peartree with
foetid blossoms
sways its high topbranches
with contrary motions and green
has come out of the wood
upon them also –

The mathematics grow complex:
there are both pinkflowered
and coralflowered peachtrees
in the bare chickenyard
of the old negro
with white hair who hides
poisoned fish-heads
here and there
where stray cats find them –
find them – find them.

O spring days, swift
and mutable, wind blowing
four ways, hot and cold.
Now the northeast wind,
moving in fogs, leaves the grass
cold and dripping. The night
is dark but in the night
the southeast wind approaches.
It is April and Einstein!
The owner of the orchard

lies in his bed
with the windows wide
and throws off his covers
one by one.

It is Einstein
out of complicated mathematics
among the daffodils –
spring winds blowing
four ways, hot and cold,
shaking the flowers!

Original publication 1921. From *Collected Earlier Poems*, copyright 1938 by New Directions Publishing Corporation, reprinted by permission of New Directions.

NOTES TO THE TEXT

Introduction

1. Wyndham Lewis, *Time and Western Man* (1927; rpt. Boston: Beacon Press, 1957), p. 126.
2. Bennett Cerf, *Out on a Limerick* (New York: Harper and Bros., 1960), p. 76. [The limerick has been attributed to Einstein himself].
3. William Carlos Williams, "St. Francis Einstein of the Daffodils," *Contact* IV (1921), p. 2.
4. Lewis, p. 212.
5. For structuralism and Foucault, see Denis Donoghue, *The Sovereign Ghost: Studies in the Imagination* (Berkeley, 1977); Michel Foucault, *The Order of Things* (New York: Pantheon, 1970); Jean Piaget, *Structuralism* (New York: Basic Books, 1970); Jacques Derrida (see recent translation of *De la grammatalogic* (Paris, 1967); Claude Levi-Strauss, *Structural Anthropology* (London and New York, 1963); Edith Kurzweil, *The Age of Structuralism: Levi-Strauss to Foucault* (New York: Columbia University Press, 1980); Robert Scholes, *Structuralism in Literature* (Yale University Press, 1974).
6. Herbert J. Muller, *Science and Criticism: The Humanistic Tradition in Contemporary Thought* (New Haven: Yale University Press, 1943), p. 92.

1 The popularization of the new physical ideas

1. A detailed critique of Jeans' and Eddington's popularizations is L. Susan Stebbing, *Philosophy and the Physicists* (London: Methuen, 1937). One of the more accurate popularizers was Edwin E. Slosson, whose impact has been carefully analyzed by David J. Rhees, in "A New Voice in Science: Science Service under Edwin E. Slosson, 1921-29," unpublished thesis, 1979, University of North Carolina, Chapel Hill.
2. "The Fabric of the Universe," *The Times* (London), 7 Nov. 1919, p. 13.
3. *The New York Times*, 9 Nov. 1919, p. 6.
4. *Ibid.*
5. *The New York Times*, 10 Nov. 1919, p. 17.
6. *The New York Times*, 11 Nov. 1919, p. 12.

7. *The Times* (London), 14 Nov. 1919, p. 8.

8. Max Born, *Physics in My Generation* (London: Methuen, 1956), p. 158.

9. See *The Times* (London), 28 Nov. 1919, p. 13 and *New York Times*, 3 Dec. 1919, p. 19.

10. *The New York Times*, 4 Dec. 1919, p. 19.

11. The limerick is published with a note on its origin in Clifton Fadiman, *The Mathematical Magpie* (New York: Simon and Schuster, 1962), pp. 289-90.

12. *The New York Times*, 7 April 1921, p. 14.

13. Bertrand Russell, *The Autobiography: The Middle Years* (Boston: Little Brown, 1967), p. 130.

14. Ronald Clark, *Einstein: The Life and Times* (New York: World Publishing, 1971), p. 247.

15. Articles on the new physics soon appeared in a wide range of journals: *The Atlantic Monthly, The Dial, The Literary Digest, Contemporary Review, The New Republic, Nature, Scientific American, Westminster Gazette, English Review, Our World, Yale Review, Quarterly Review, Fortnightly Review, Mind, Science, Edinburgh Review, Dublin Review, The Monist, North American Review, Popular Astronomy, The Nation, The New Statesman, Harper's Magazine, American Journal of Psychology, Philosophical Review, Forum, Living Age, Discovery* – to mention only some of the popular magazines and non-physics journals which spread the word to their English-speaking readers.

16. *The New Republic*, 18 February 1920, p. 345. The articles were in the 21 January issue, pp. 228-31, and the 18 February issue, pp. 341-5.

17. *The Nation*, 27 Dec. 1919, p. 819.

18. *The Nation*, 7 April 1920, p. 503.

19. *The Atlantic* 126, Oct. 1920, p. 570.

20. *The Nation*, 3 June 1925, p. 619.

21. *The Nation*, 10 June 1925, p. 652.

22. *The Nation*, 17 June 1925, p. 686.

23. *The Nation*, 24 June 1925, p. 712.

24. *The Times* (London), 3 April 1922, p. 13.

25. Richard Feynman, *The Feynman Lectures in Physics* (Reading, Mass.: Addison-Wesley Publishing Co., 1965), vol. I, chapter 16, p. 1.

26. Clark, p. 314.

27. A few of these works (written mostly by physicists, mathematicians, and philosophers) include the following: Erwin Freundlich, *The Foundations of Einstein's Theory of Gravitation* (Cambridge University Press, 1920); R.B. Haldane, *The Reign of Relativity* (London: J. Murray, 1921); J.H. Thirring, *The Ideas of Einstein's Theory*, trans. Rhoda A.B. Russell (London: Methuen, 1921); H. Weyl, *Space-Time-Matter*, trans. H.L. Brose (London: Methuen, 1922); H. Weyl, *The Open World* (New Haven: Yale University Press and London: Oxford University Press, 1932); Max Planck, *The Origin and Development of the Quantum Theory* (Oxford University Press, 1922); Charles Nordmann, *Einstein and the Universe*, trans. Joseph McCabe (London and New York: Henry Holt and Co., 1922); H.A. Lorentz, A. Einstein, H. Minkowski, and H. Weyl, *The*

Principle of Relativity: A Collection of Original Memoirs on the Special and General Theory of Relativity, trans. W. Perrett and G.B. Jeffrey (New York: Dodd, Mead and Co. and London: Methuen, 1923); Max Born, *Einstein's Theory of Relativity*, trans. H.L. Brose (London: Methuen, 1924); Max Planck, *A Survey of Physics*, trans. R. Jones and D.H. Williams (London: Methuen, 1925); Charles Nordmann, *The Tyranny of Time: Einstein or Bergson*, trans. J. McCabe (London and New York: H. Holt and Co., 1925); George David Birkhoff, *Relativity and Modern Physics*, (Cambridge, Mass.: Harvard University Press, 1923); Charles D. Broad, *Scientific Thought* (London: Methuen, 1922); Samuel Dingle, *Relativity for All* (London: Methuen, 1922); Samuel Gugenheimer, *The Einstein Theory Explained and Analyzed* (New York: Macmillan, 1925); Benjamin Harrow, *From Newton to Einstein* (New York: D. Van Nostrand, 1920); Archibald Henderson, *Relativity, a Romance of Science* (Chapel Hill, North Carolina: University of North Carolina Press, 1923); William Franklin Hudgings, *Introduction to Einstein and Universal Relativity* (New York: Arrow Book Co., 1922); James Rice, *Relativity: A Systematic Treatment of Einstein's Theory* (London and New York: Longmans, Green and Co., 1923); E.E. Slosson, *Easy Lessons in Einstein* (New York: Harcourt, Brace and Howe, 1920).

28. Clark, p. 313.
29. *The New York Times*, 3 April 1921, p. 1.
30. *Literary Digest*, 16 April 1921, p. 34.
31. *The New York Times*, 27 April 1921, p. 21.
32. *The Plain Dealer*, 26 May 1921, p. 1.
33. *Manchester Guardian*, 10 June 1921.
34. See *The Times* (London), 17 June 1921, p. 8; G.K. Bell, *Randall Davidson, Archbishop of Canterbury* (London, 1935), p. 1052; Lord Rayleigh, *The Life of Sir J.J. Thomson OM*. (Cambridge, 1942), p. 203; Carl Seelig, *Albert Einstein: A Documentary Biography* (London, 1956), p. 79.
35. Michel Biezunski, "Dossier: Einstein à Paris," *La Recherche* 132, April 1982, pp. 502-10.
36. Clark, p. 246.
37. Pablo Picasso, "Art as Individual Idea," in *The Modern Tradition*, ed. Richard Ellmann and Charles Feidelson (New York: Oxford University Press, 1965), p. 26.
38. Guillaume Apollinaire, "Pure Painting," *The Modern Tradition*, p. 116.
39. Maurice Raynal, *The Skira Modern Painting*, trans. Stuart Gilbert (Switzerland: World Publishing, 1960), p. 93.
40. *Dial* 70, May 1921, pp. 535-6.
41. Jose Ortega y Gasset, *The Modern Theme*, trans. James Clough (New York: Harper and Bros., 1916), p. 136.
42. John A. Richardson, *Modern Art and Scientific Thought* (Urbana, Illinois: University of Illinois Press, 1971), p. 105.
43. Daniel-Henry Kahnweiler, *Juan Gris: His Life and Work*, trans. Douglas Cooper (New York: Henry N. Abrams, 1946), p. 31.
44. Kahnweiler, pp. 194, 199, 202.

45. Kahnweiler, p. 178.
46. Kahnweiler, p. 190.
47. Sigfried Giedion, *Space, Time and Architecture* (Cambridge, Mass.: Harvard University Press, 1967), p. 366.
48. Jose Ortega y Gasset, *The Dehumanization of Art and Notes on the Novel*, trans. Helen Weyl (Princeton University Press, 1948), p. 9.
49. Virginia Woolf, *Roger Fry* (New York: Harcourt, Brace and Co., 1940), p. 153.
50. *1913 Armory Show 50th Anniversary Exhibition: 1963* (New York: Henry Street Settlement and Utica, New York: Munson-Williams-Proctor Institute, 1963), p. 35.

2 Newtonian mechanics and literary responses

1. William Shakespeare, *Troilus and Cressida* (1602), Act I, Scene III, lines 85-103.
2. A challenging and fascinating argument for these connections is made by E.M.W. Tillyard, *The Elizabethan World Picture* (New York: Macmillan Press, 1952). Background information is provided by Arthur O. Lovejoy, *The Great Chain of Being* (Cambridge, Mass.: Harvard University Press, 1936), and Marie Boas, *The Scientific Renaissance 1450-1630* (New York: Harper and Brothers, 1962).
3. Robert Karplus, *Introductory Physics: A Model Approach* (New York: W. A. Benjamin, 1969), p. 57.
4. The law is expressed by the formula $F = GMm/r^2$ and states that two bodies attract each other with a force (F) that is proportional to the product of the masses of each body (M and m) and is inversely proportional to the square of the distance (r) between them. G is a fixed number, a constant. For an engaging discussion of this principle, and all of Newtonian physics, see Eric Rogers, *Physics for the Inquiring Mind* (Princeton University Press, 1960).
5. John Donne, "An Anatomy of the World: The First Anniversary" (1611), lines 205-18.
6. *The Clockwork Universe: German Clocks and Automata 1550-1650*, ed. Klaus Maurice and Otto Mayr (New York: Neal Watson Academic Publications, 1980).
7. A. d'Abro, *The Rise of the New Physics* (1939; rpt. New York: Dover, 1951), p. 104.
8. Arthur Hugh Clough, "The New Sinai," in *The Poems of Arthur Hugh Clough*, 2nd ed., F.L. Mulhauser (1849; rpt. Oxford University Press, 1974).
9. Marjorie Nicolson, *Newton Demands the Muse* (Princeton University Press, 1946) and Douglas Bush, *Science and English Poetry* (New York: Oxford University Press, 1950).
10. Nicolson, pp. 22-23.
11. James Thomson, "To the Memory of Newton," in *The Complete Poetical Works of James Thomson*, ed. J.L. Robertson (London and New York: Oxford University Press, 1908), pp. 436-42.

12. Nicolson, p. 70.
13. Bush, note 9, p. 53.
14. James Thomson, pp. 437, 438.
15. Alexander Pope, "Essay on Man," II, lines 19-30.
16. Alexander Pope, "The Dunciad," IV, lines 471-8.
17. Alexander Pope, "The Dunciad," III, line 217.
18. Keats, *Lamia*, Part II, lines 229-37.
19. Coleridge, "The Theory of Life," from *Hints Toward a More Comprehensive Theory of Life*, in *Selected Poetry and Prose of Samuel Taylor Coleridge*, ed. Donald Stauffer (New York, Random House: 1951), lines 229-237.
20. Edward Proffit, "Science and Romanticism," *The Georgia Review* XXXIV (Spring 1980), p. 61.
21. Chapter XVIII, *Biographia Literaria* (London and New York: Dutton and Co., 1906), p. 208.
22. Bush, p. 152.
23. Whitman, "To a Locomotive in Winter," *The Norton Anthology of Poetry*, ed. Eastman *et al.* (New York: Norton, 1970), p. 813.
24. Dickinson, "I Like to See It Lap the Miles," *Ibid.*, p. 850.
25. George Becker, ed., *Documents of Modern Literary Realism* (Princeton University Press, 1967), p. 34.
26. Emile Zola, in Becker, p. 172.
27. Zola, p. 166.
28. Zola, p. 208.
29. Gustave Flaubert, "On Realism" in Becker, p. 95.
30. Crane, "War is Kind," XXI, *The Collected Poems of Stephen Crane* (New York: Alfred A. Knopf, 1922), p. 101.
31. Walt Whitman, "When I Heard the Learn'd Astronomer," *Leaves of Grass*, ed. Sculley Bradley and Harold Blodgett (New York: W. W. Norton and Co., 1973), p. 271.
32. G.H.A. Cole, "Physics," in *The Twentieth Century Mind*, II, ed. C. B. Cox and A. E. Dyson (Oxford University Press, 1972), p. 229.
33. Stanley Jaki, *The Relevance of Physics* (University of Chicago Press, 1966), pp. 481, 482.
34. Jaki, p. 484.
35. Maxwell, "A Dynamical Theory of the Electromagnetic Field," *Scientific Papers* (Cambridge University Press, 1890), vol. I, as reprinted in *Physical Thought/from the Presocratics to the Quantum Physicists*, ed. Shmuel Sambursky (New York: Pica Press, 1974) p. 437.
36. Francis Thompson, "The Mistress of Vision," in *Complete Poetical Works of Francis Thompson* (New York, Modern Library, n.d. [1903?]), p. 184.

3 Einstein's revolution

1. Max Born, *Einstein's Theory of Relativity* (1924; rpt. New York: Dover Publishing, 1962), p. 86.
2. Maxwell, "A Dynamical Theory of the Electromagnetic Field," *Scientific Papers* (Cambridge University Press, 1890), vol. II, as reprinted in *Physical Thought/from the Presocratics to the Quantum Physicists*, ed. Shmuel Sambursky (New York: Pica Press, 1974) pp. 443-4.

3. Stanley Jaki, *The Relevance of Physics* (University of Chicago Press, 1966), p. 84.
4. Jaki, p. 82.
5. In metric units, about 300,000,000 meters per second (best current value: 299,792,458). We use British units for the speed of light because the figure of 186,000 miles per second is so familiar, and for non-scientist readers we do not wish to add the burden of learning metric units to that of learning about relativity. We do note, however, that metric units are universally used by contemporary scientists. All references in this book to the speed of light mean the speed of light *in a vacuum*. Light in air travels very slightly slower, and light in glass or water is considerably slower.
6. For a detailed discussion, see references to Einstein and the Michelson-Morley experiment in Gerald Holton, *Thematic Origins of Scientific Thought, Kepler to Einstein* (Cambridge, Mass.: Harvard University Press, 1973); Loyd S. Swenson, Jr., *Genesis of Relativity* (New York: Burt Franklin and Co., 1979), and Abraham Pais, *"Subtle is the Lord..."/The Science and the Life of Albert Einstein* (New York: Oxford University Press, 1982).
7. "Elektrodynamik liewegter Korper," *Annalen der Physik*, ser. 4, vol. 17, pp. 891-921.
8. See Swenson, note 6.
9. See Robert E. Gibbs, "Photographing a Relativistic Meter Stick," *American Journal of Physics*, 48/12 (1980), pp. 1056-8, and G.D. Scott and M.R. Viner, "The Geometrical Appearance of Large Objects Moving at Relativistic Speeds," *American Journal of Physics*, 33/7 (1965), pp. 534-6.
10. That is a possibility in contemporary cosmology. See any astronomy text after 1970, such as Joseph Silk, *The Big Bang* (San Francisco: Freeman, 1980). A good popular treatment is William Kaufmann, *Cosmic Frontiers of General Relativity* (San Francisco: Freeman, 1977). The verdict on whether this possibility is realized depends on measurements still underway. Accurate review articles for lay readers appear periodically in the popular-level journal *Mercury* (Astronomical Society of the Pacific, 1290 24th Avenue, San Francisco, CA 94122).
11. Isaac Newton, *Mathematical Principles,* 1686, trans. and ed. Florian Cajori (Berkeley: University of California Press, 1946), p. 6.
12. Albert Einstein, "Autobiographical Notes," in *Albert Einstein: Philosopher-Scientist*, 2nd ed., ed. Paul A. Schilpp (New York: Tudor Publishing, 1951), p. 27.
13. Although 1915 is commonly given as the date for the General Theory, it emerged from a series of papers between 1910 and 1916, culminating in a complete presentation as "Grundlage der allgemeinen Relativitats-theorie," *Annalen der Physik*, ser. 4, vol. 49 (1916), pp. 769-822.
14. There is no evidence that Galileo actually did the experiment at the tower of Pisa as legend would have it, but he observed this equivalence in many different experiments.
15. Einstein, p. 65. Note that "acceleration" is used in physics to describe any change in velocity, whether that change is an increase or decrease in speed,

or a change in direction. By contrast, the common usage of the word applies only to an increase in speed.

16. Born, p. 318.
17. "According to the general theory...not only should Mercury travel around the sun, but the ellipse which it describes should rotate very slowly relative to the CS [coordinate system] connected with the sun. This rotation of the ellipse expresses the new effect of the general relativity theory. The new theory predicts the magnitude of this effect. Mercury's ellipse would perform a complete rotation in three million years." Einstein, "The Meaning of Relativity," in *Great Ideas in Modern Science*, ed. Robert Marks (New York: Bantam Books, 1967), p. 165.
18. If light is assumed to be composed of particles (as Newton believed) and these particles are assumed to have non-zero mass, Newtonian physics would also predict some bending of light rays, but not the same amount of bending as in Einstein's curved space-time theory. The actual bending observed is Einstein's value, however, and rules out a Newtonian model.
19. Louis de Broglie, "A General Survey of the Scientific Work of Albert Einstein," in Schilpp (note 12), p. 119.
20. Einstein, in *Great Ideas in Modern Science*, (note 17) pp. 162, 164. See the references in note 10 for current treatments of these ideas.
21. Eric Rogers (personal communication, 1984).
22. Richard Feynman, *The Feynman Lectures on Physics* (Reading, MA: Addison-Wesley, 1965) vol. I, chapter 20, p. 6.
23. J. Willard Gibbs, quoted in Stanley Jaki, *The Relevance of Physics* (University of Chicago Press, 1966), p. 118.

4 Einstein becomes a muse

1. *Current Opinion*, January 1920, pp. 72-3.
2. *Harpers*, March 1920, pp. 477-87.
3. *The New Republic*, 21 January 1920 and 18 February 1920.
4. *The New Republic*, 6 July 1921, pp. 172-4.
5. William Carlos Williams, "St. Francis Einstein of the Daffodils," *Contact* IV (1921), pp. 2-4.
6. Williams, "On the first visit of Professor Einstein to the United States in the spring of 1921," *Adam & Eve & the City*, in *Collected Later Poems* (New York: New Directions, 1962).
7. Williams, note 5.
8. Thomas Jewell Craven, "Art and Relativity," *The Dial* 70 (May 1921), pp. 535-9.
9. Williams, "The Poem as a Field of Action," *Selected Essays of William Carlos Williams* (New York: Random House, 1954), p. 283.
10. Williams, "On Measure – Statement for Cid Corman," *Selected Essays of William Carlos Williams* (New York: Random House, 1954), pp. 337, 340.
11. Williams, *Paterson* (New York: New Directions, 1936), p. 50.
12. Kay Boyle and Robert McAlmon, *Being Geniuses Together* (New York: Doubleday, 1968), p. 258.

13. Archibald MacLeish, "Einstein," in *FIets in the Moon* (Boston: Houghton Mifflin, 1926). Also in *Poems: 1924-33* (Boston: Houghton Mifflin, 1933), pp. 67-75.
14. Hyatt Howe Waggoner, *The Heel of Elohim: Science and Values in Modern American Poetry* (Norman: University of Oklahoma Press, 1950), pp. 143-4.
15. Anton Reiser, *Albert Einstein* (London: Butterworth, 1931).
16. *Prepositions: The Collected Critical Essays of Louis Zukofsky* (New York: Horizon Press, 1968), p. 7.
17. Zukofsky, p. 51.
18. *Literature: An Introduction to Fiction, Poetry and Drama* (Boston: Little Brown and Co., 1976), p. 617.
19. Zukofsky, *"A": 1-12* (Garden City, New York: Doubleday, 1967), pp. 144, 179.
20. Reiser, p. 197.
21. *Albert Einstein: The Human Side*, ed. Helen Dukas and Banesh Hoffmann (Princeton University Press, 1979), p. 76.
22. Reiser, p. 190.
23. Olson, "Equal, That is, to the Real Itself," *Poetics of the New American Poetry*, ed. Donald Allen and Warren Tallman (New York: Grove Press, 1973), pp. 177-8.
24. Olson, p. 181.
25. Robert Frost, *The Poetry of Robert Frost* (New York: Holt, Rinehart and Winston, 1969), p. 396.
26. Lawrance Thompson, *Robert Frost: The Years of Triumph, 1915-1938* (New York: Holt, Rinehart and Winston, 1970), p. 288.
27. Thompson, p. 658.
28. Frost, p. 389.
29. David D. Pearlman, *The Barb of Time: On the Unity of Ezra Pound's Cantos* (New York: Oxford University Press, 1969), p. 204.
30. Ezra Pound, *Guide to Kulchur* (London: Faber & Faber, 1938), p. 34; rpt. 1952, New Directions.
31. T.S. Eliot, "Thoughts After Lambeth," *Selected Essays* (New York: Harcourt, Brace and World, 1950), p. 327.
32. Eliot, p. 328.
33. "Nature of Space: Professor Einstein's Change of Mind," *The Times* (London), 6 Feb. 1931.
34. T.S. Eliot, "The Waste Land," (1922) rpt. in *The Complete Poems and Plays 1909-1950* (New York: Harcourt, Brace and World, 1971) p. 50, lines 427, 431.
35. "The Hollow Men," (1925), in ref. 34, p. 58.
36. "The Waste Land," in ref. 34, p. 38, line 20.
37. "The Rock," in ref. 34, p. 104.
38. E. E. Cummings, letter to Kenneth Burke, in *Selected Letters of E. E. Cummings* (New York: Harcourt, Brace and World, 1969), p. 248.
39. Letter to Eva Hesse, in *Selected Letters*, p. 265.

40. E.E. Cummings, "Space being(don't forget to remember)Curved," in *Norton Anthology of Modern Poetry*, ed. Ellmann and O'Clair (New York: Harcourt Brace Jovanovich, 1972), p.315.
41. Karel Čapek, *Krakatit*, 1924, trans. Lawrence Hyde (New York: Macmillan, 1925).
42. *Ibid.*, p. 19.
43. *Ibid.*, p. 348.
44. Wallace Stevens, "Thirteen Ways of Looking at a Blackbird," *The Collected Poems* (New York: Vintage Books, 1982), pp. 92-5.
45. Frederick L. Gwynn and Joseph Blotner, eds., *Faulkner in the University* (Charlottesville: University of Virginia Press, 1959), pp. 273-4.
46. Gwynn and Blotner, p. 139.
47. Virginia Woolf, "The Sentimental Journey," *Collected Essays* vol. I (New York: Harcourt, Brace and World, 1978), p. 97.
48. Joseph Frank, "Spatial Form in Modern Literature," *The Widening Gyre* (Bloomington: Indiana University Press, 1963), p. 13.
49. Lawrence Durrell, *Balthazar* (New York: E. P. Dutton, 1958), p. 5.
50. Durrell, *A Key to Modern British Poetry* (1952; reprint Norman: University of Oklahoma Press, 1970), p.26.
51. Durrell, *Balthazar*, p. 142.
52. Durrell, *Justine* (New York: E.P. Dutton, 1957), p. 27.
53. Durrell, *Clea* (New York: E.P. Dutton, 1960), p. 135.
54. Jacob Bronowski, *Insight – Ideas of Modern Science* (New York: Harper and Row, 1964), p. 107.
55. Parts of this section are based on a talk by Alan Friedman presented at the 1973 Northeast Modern Language Association meeting. That talk was also the basis for an article, "The Novelist and Modern Physics: New Metaphors for Traditional Themes," *Journal of College Science Teaching* 4/5 (1975), pp. 310-2.
56. Alfred Appel, "An Interview with Nabokov," *Wisconsin Studies in Contemporary Literature*, 8 (Spring 1967), pp. 140-1.
57. Vladimir Nabokov, *Transparent Things* (New York: McGraw Hill, 1972), pp. 13-14.
58. For example, Michael Rosenblum, "Finding What the Sailor has Hidden: Narrative as Patternmaking in *Transparent Things*," *Contemporary Literature* XIX, no. 2, (Spring 1978), pp. 219-32. The remark about physics cited in note 6 of this paper was made by one of the present authors (AJF) to Rosenblum at the December 1976 Modern Language Association Meeting.
59. Albert Einstein, trans. Robert W. Sheffield, *Relativity/The Special and the General Theory* (1916; 5th ed. rpt. New York: Crown Publishers [n.d.]), p. 25.
60. Nabokov has admonished authors to enclose the word in quotations.
61. Vladimir Nabokov, *Ada or Ardor: A Family Chronicle* (New York: McGraw Hill, 1969), p. 539. Page numbers following are to this edition.

62. Alfred Appel, "Ada," *New York Times Book Review*, 4 May 1969, p. 34.
63. Martin Gardner, *The Ambidextrous Universe* (1964; rpt. New York: Charles Scribner's Sons, 1979), p. 152.
64. Virginia Woolf, *Collected Essays* vol. I (New York: Harcourt, Brace and World, 1967), pp. 320-1.
65. Woolf, "How It Strikes a Contemporary," *Collected Essays* vol. II, pp. 157-8.
66. William James, *A Pluralistic Universe* (London: Longmans, Green, 1909), p. 254.
67. Woolf, *Jacob's Room* and *The Waves* (1922 and 1931; rpt. New York: Harcourt, Brace and World, 1959), p. 189. All further references noted parenthetically.
68. William Faulkner, *The Sound and The Fury* (1929; rpt. New York: Vintage Books, 1946), pp. 97, 94. All further references noted parenthetically.
69. Faulkner, *Absalom, Absalom!* (1936; rpt. New York: The Modern Library, 1951), p. 101. All further references noted parenthetically.
70. Gwynn and Blotner, p. 151.
71. Gwynn and Blotner, pp. 41-42.
72. Faulkner, *As I Lay Dying* (1930; rpt. New York: Vintage Books, 1957), p. 139.
73. Faulkner, *Light in August* (1932; rpt. New York: Modern Library, 1968), pp. 345-6. All further references noted parenthetically.
74. Part of the following section on *Ulysses* is taken from an oral presentation by one of the authors (Alan J. Friedman) for the 1979 International James Joyce Society meeting in Zurich. This talk was also the basis for an article in *The Seventh of Joyce*, ed. Bernard Benstock (Bloomington: Indiana University Press, and Sussex: Harvester Press, 1982) pp. 198-206.
75. Richard Ellmann, *James Joyce* (1959; rpt. Oxford University Press, 1976), p. 462.
76. Ronald Clark, *Einstein: The Life and Times* (New York: World Pub., 1971), p. 220.
77. Ellmann, p. 537.
78. Ellmann, p. 515.
79. James Joyce, *Letters of James Joyce*, ed. Stuart Gilbert (New York: Viking Press, 1957), p. 164.
80. Joyce, *Letters*, p. 178.
81. Richard Kain, *Fabulous Voyager* (New York: Viking Press, 1959), pp. 185-8. See also comments on science in *Ulysses* on pp. 9, 22, 227.
82. Kain, pp. 228, 233, 236.
83. Mark Littmann and Charles Schweighauser, "Astronomical Allusions, Their Meaning and Purpose, in Ulysses," *James Joyce Quarterly*, II (1965), pp. 238-46.
84. Wyndham Lewis, *Time and Western Man* (1927; rpt. Boston: Beacon Press, 1957), p. 103.
85. Avron Fleishman, "Science in Ithaca," *Wisconsin Studies in Contemporary Literature*, VII (1967), pp. 377-91.
86. Fleishman, p. 390.

87. Fleishman, p. 381.
88. W. Y. Tindall, *James Joyce: His Way of Interpreting the World* (New York: Charles Scribner's Sons, 1950), p. 90.
89. Edward Watson, "STOOM-BLOOM: Scientific Objectivity versus Romantic Subjectivity in the Ithaca Episode of Joyce's *Ulysses*," *University of Windsor Review*, II (1966), pp. 11-25.
90. See, for example, Jacob Bronowski, *A Sense of the Future* (Cambridge, Mass.: MIT Press, 1977); Gerald Holton, *Thematic Origins of Scientific Thought: Kepler to Einstein* (Cambridge, Mass.: Harvard University Press, 1973); Thomas Kuhn, *The Structure of Scientific Revolutions*, 2nd ed., rev. (University of Chicago, 1970).
91. William York Tindall, *A Reader's Guide to James Joyce* (New York: The Noonday Press, 1959), pp. 247-8.
92. James Joyce, *Finnegans Wake* (New York: Viking Press, 1939), p. 149. All further references will be noted in the text.
93. Clive Hart, *Structure and Motif in Finnegans Wake*, (Northwestern University Press, 1962), pp. 65-6.
94. Tony Tanner, *City of Words* (New York: Harper and Row, 1971), p. 21.
95. Sharon Spencer, *Space, Time and Structure in the Modern Novel* (New York University Press, 1971), chapters 8 and 9.
96. James Jeans' *The Mysterious Universe* (see note 1, chapter 1), along with other popularizations, is cited repeatedly in Durrell's own criticism (note 50, above). Nabokov quotes popularizer Martin Gardner's *The Ambidextrous Universe* (note 63, above) and much of *Ada*'s twin imagery from science probably came from Gardner. Nabokov's treatments of relativity are very close to another Gardner popularization, *Relativity for the Millions* (New York: Macmillan, 1962).

5 The second revolution

1. Robert March, *Physics for Poets* (New York: McGraw-Hill, 1970), p. 199. This is one recommended non-mathematical introduction to contemporary physics, including quantum theory. Another is Adolph Baker, *Modern Physics and Antiphysics* (Reading, MA: Addison-Wesley, 1970). Two recent popularizations provide the first detailed treatments of the quantum theory including its various interpretations. Fred Alan Woolf's *Taking the Quantum Leap* (San Francisco: Harper and Row, 1982) is the most comprehensive, and makes for difficult but rewarding reading. Paul Davies' *Other Worlds* (New York: Simon and Schuster, 1980) is shorter and less thorough.
2. Max Planck, *A Survey of Physical Theory*, trans. R. Jones and D.H. Williams (1925; rpt. New York: Dover Publications, 1960), pp. 108-9. [Formerly titled: *A Survey of Physics*].
3. Gerald Holton, *Thematic Origins of Scientific Thought: Kepler to Einstein* (Cambridge: Harvard University Press, 1973) [citation Miller's].
4. Arthur I. Miller, "Visualization Lost and Regained: The Genesis of the Quantum Theory in the Period 1913-27," in *On Aesthetics in Science*, ed. J.

Wechsler (Cambridge, Mass.: Massachusetts Institute of Technology Press, 1978) p. 75.

5. A summary of the various versions and origins of this phrase appear in Ronald Clark, *Einstein: The Life and Times* (New York: World Publishing, 1971), pp. 340-5. See also note 10, below.

6. Einstein letter, 29 April 1924 in *The Born-Einstein Letters*, ed. Max Born, trans. Irene Born (London: Macmillan, 1971), p. 82.

7. D'Abro, *The Rise of the New Physics* (1939; rpt. New York: Dover Publications, 1951), p. 944.

8. See Woolf (note 1), pp. 202ff. For a much more technical treatment, see Bernard d'Espagnat, "The Quantum Theory and Reality," *Scientific research*, 241/5 (November 1979), pp. 158-81.

9. d'Espagnat (note 8) offers a version of what various resolutions to the E.P.R. experiment will mean for the interpretation of quantum theory, but his conclusions are not universally accepted by physicists. A running debate is conducted in professional journals, the most accessible of which is *Physics Today*. Periodic popular level summaries appear in *New Scientist*.

10. A full treatment of Einstein's contributions to and dissatisfactions with the quantum theory is provided in Abraham Pais, *"Subtle is the Lord..."/The Science and the Life of Albert Einstein* (New York: Oxford University Press, 1982) Section VI.

11. James Jeans, *Physics and Philosophy* (Cambridge University Press, 1943), p. 202.

12. Weyl, *Philosophy of Mathematics and Natural Science* (Princeton University Press, 1949), p. 116.

13. Louis de Broglie, *Physics and Light: The New Physics*, trans. W.H. Johnston (New York: Norton, 1939), p. 252.

14. Herbert J. Muller, *Science and Criticism: The Humanistic Tradition in Contemporary Thought* (New Haven: Yale University Press, 1964), p. 78.

15. Muller, p. 79.

16. James, "The Dilemma of Determinism," *The Writings of William James*, ed. John J. McDermott (New York: Random House, 1967), pp. 593, 599.

17. Milič Čapek, "Bergson and Modern Physics," *Boston Studies in the Philosophy of Science*, VII (New York: Humanities Press, 1971), p. 90.

18. Bergson, *Time and Free Will*, trans. F.L. Pogson (1913; rpt. New York: Harper, 1960), p. 212.

19. Whitehead, *Science and the Modern World*, (1925; rpt. New York: The Free Press, 1967), p. 153.

20. Whitehead, *Essays in Science and Philosophy* (New York: Philosophical Library, 1948), p. 90.

21. Russell, *ABC of Relativity* (1925; rpt. New York: Mentor Books, 1969), p. 141.

22. Russell, pp. 135-6.

23. Quoted in Stanley Jaki, *The Relevance of Physics* (University of Chicago Press, 1966), p. 570.

24. Jaki, pp. 362-3.

25. Jaki, p. 363.
26. Jeans, *Physics and Philosophy*, p. 216.
27. A.S. Eddington, *The Nature of the Physical World* (Cambridge University Press, 1928), p. 295.
28. Clark, p. 275.
29. Henri Poincaré, *The Foundation of Science* (Lancaster, Penna.: Science Press, 1913), p. 206.
30. Heisenberg, *Physics and Philosophy: The Revolution in Modern Science* (New York: Harper and Row, 1958), p. 50.
31. Heisenberg, pp. 52, 58.
32. Einstein quoted in *The Saturday Evening Post*, 26 Oct. 1929, p. 17.
33. Gerald Holton, note 3, p. 96.
34. For various means by which randomness entered modern science, see Stephen G. Brush, "Irreversibility and Indeterminism: Fourier to Heisenberg," *Journal of the History of Ideas*, 37 (1976), pp. 603-30, and Alfred M. Bork, "Randomness and the Twentieth Century," *Antioch Review* XXVII (1967), pp. 40-61.
35. W.H. Auden, *The Dyer's Hand and Other Essays* (New York: Random House, 1962), p. 78.
36. W.H. Auden, *New Year Letter* (London: Faber and Faber, 1941). All further references will be noted parenthetically.
37. Auden, "Notes to Letter," in *New Year Letter*, p. 119.
38. Wallace Stevens, preface to William Carlos Williams' *Collected Poems: 1921-1931* (New York: The Objective Press, 1934), p. 2.
39. William Carlos Williams, *Paterson* (New York: New Directions, 1936), p. 176. All further references will be noted parenthetically.
40. Hans Vaihinger, *The Philosophy of 'As if '*, trans. G.K. Ogden (London: Routledge and Kegan Paul, 1949), p. 61.
41. Holton, p. 148.
42. Charles Olson, "Equal, That Is, to the Real Itself," *Poetics of the New American Poetry*, ed. Donald Allen and Warren Tallman (New York: Grove Press, 1973), pp. 179-80.
43. Robert Duncan, "Ideas of the Meaning of Forms," *Poetics*, p. 207.
44. Olson, p. 181.
45. William James, *A Pluralistic Universe* (London: Longmans, Green and Co., 1909), p. 219.
46. Virginia Woolf, *A Writer's Diary*, ed. Leonard Woolf (New York: Harcourt, Brace, and Co., 1954), p. 93.
47. Quentin Bell, *Virginia Woolf: A Biography*, II (New York: Harcourt, Brace, Jovanovich, 1972), p. 106.
48. Woolf, note 46, p. 136.
49. Nabokov's *Ada* is another example, as discussed earlier (p. 92).
50. Samuel Beckett, *Waiting for Godot* (New York: Grove Press, 1954), p. 27.
51. Beckett, p. 35.
52. Martin Esslin, *The Theatre of the Absurd* (New York: Doubleday, 1969), p. 25.

53. Virginia Woolf, *Jacob's Room* and *The Waves* (1922 and 1931; rpt. New York: Harcourt, Brace and World, 1959), p. 322. All further references will be noted parenthetically.

54. Woolf, note 46, p. 143.

55. Richard Huelsenbeck, "en Avant Dada," *The Dada Painters and Poets*, ed. Robert Motherwell (New York: Wittenborn, Schultz, Inc., 1951), p. 29.

56. Jean (Hans) Arp, "Dada Was Not a Farce," in *The Dada Painters and Poets*, p. 294.

57. Hugo Ball, "Dada Fragments," *The Dada Painters and Poets*, pp. 51, 53.

58. Tom Stoppard, *Travesties* (New York: Grove Press, 1975), p. 37.

59. *Ibid.*, p. 53.

60. Martin Heidegger, "Dread Reveals Nothing," *The Modern Tradition*, ed. Richard Ellmann and Charles Feidelson (New York: Oxford University Press, 1965), p. 838.

61. Ernest Hemingway, "A Clean, Well-lighted Place," *Snows of Kilimanjaro and Other Stories* (New York: Scribner, 1961), pp. 32-3.

62. Jean-Paul Sartre, "Choice in a World Without God," *The Modern Tradition*, note 60, p. 837.

63. Sartre, "Authenticity," *The Modern Tradition*, note 60, p. 842.

64. Albert Camus, "The Fact of Absurdity," *The Modern Tradition*, note 60, pp. 826-7.

65. Camus, "Absurd Freedom," *The Modern Tradition*, note 60, p. 852.

66. Julio Cortázar, "Blow-up," in *Blow-Up and Other Stories* (New York: Collier Books, 1967).

67. Sharon Spencer, *Space, Time and Structure in the Modern Novel* (New York University Press, 1971), p. 57.

68. William Faulkner, *Absalom, Absalom!* (1936; rpt. New York: The Modern Library, 1951), p. 217. All further references will be noted parenthetically.

69. James Guetti, "*Absalom, Absalom!*: The Extended Simile," in *Twentieth Century Interpretations of Absalom, Absalom!*, ed. Arnold Goldman (Englewood Cliffs, New Jersey: Prentice-Hall, 1971), p. 97.

70. Anthony Burgess, *ReJoyce* (New York: W. W. Norton, 1965), pp. 222-223.

71. James Joyce, *Finnegans Wake* (New York: Viking, 1939) p. 149. All further references will be noted parenthetically.

72. John Graham, "Time in the Novels of Virginia Woolf," *Critics on Virginia Woolf*, ed. Jacqueline Latham (London: George Allen and Unwin, 1970).

73. Arthur Eddington, *The Philosophy of Physical Science* (New York: Macmillan, 1939), pp. 110-11.

74. Woolf, *Collected Essays* IV (New York: Harcourt, Brace and World, 1967), p. 182.

75. Frederick Gwynn and Joseph Blotner, eds., *Faulkner in the University* (Charlottesville: University of Virginia Press, 1959), pp. 38-9.

76. Gwynn and Blotner, p. 39.

77. Esslin, pp. 63-4.

78. Robert Coover, *The Universal Baseball Association, J. Henry Waugh, Prop.*, (New York: Random House, 1968). Page references are to the

paperback edition (New York: Signet, 1969). The first suggestion to the present authors that Coover might be using concepts from quantum theory came from Mickey Friedman (Mrs. Alan Friedman) in 1970. This discussion is based largely on a paper by Alan J. Friedman published in *Trema* (University of Paris III), No. 1, 1976, pp. 147-54. Professor Arlen Hansen had independently considered this possibility (see next note), and we enjoyed discussions before either of our publications appeared.

79. Arlen Hansen, *Novel* 10 (Fall 1976), pp. 54-5.
80. Robert Coover (personal communication to Alan J. Friedman, 1973).

6   A myth portrayed

1. *Time* cover by Ernest Tamlin Baker, XLVIII, no. 1 (1 July 1946).
2. Albert Einstein, "Ist die Tragheit eins Korpers von seinem Energieinhalt abhangig?" *Annalen der Physik*, ser. 4, vol. 18, (1905) pp. 639-41.
3. Albert Einstein, "Uber das Relativitatsprinzip und die aus demselben gezogenen Flogerungen," *Jahrbuch der Radioactivitat und Elektronik*, vol. 4. (1907), pp. 411-62, and vol. 5. (1908), pp. 98-9 (Berichtigungen).
4. For example, Edwin E. Slosson, in *Easy Lessons in Einstein* (New York: Harcourt, Brace, and Howe, 1920) discusses mass-energy relations directly on only one page (96) and without printing $E = mc^2$ at all. J.H. Thirring, *The Ideas of Einstein's Theory* (trans. Rhoda A.B. Russel; London: Methuen and Co., 1921) allots 6 pages (87-92), and describes the equation without using the now familiar symbols.
5. *Pittsburgh Post-Gazette*, December 29, 1934, Second News Section, p. 1. (1908), pp. 98-9 (Berichtigungen).
6. For a summary of this debate, see Loren Eiseley, *Darwin's Century* (New York: Doubleday and Co., 1958). Chapter IX, "Darwin and the Physicists," is most directly concerned with this topic. The history of atomic energy discussed in this section has been the subject of numerous studies. Recommended non-technical works are: Emilio Segre, *From X-Rays to Quarks* (San Francisco: W.H. Freeman, 1980; original edition, *Personaggi e Scoperte nella Fisica Contemporanea*, Mondadori, Milan, 1976) which presents the story in the context of other developments in physics; Robert Jungk, *Brighter than a Thousand Suns* (1956; trans. James Cleugh, New York: Harcourt, Brace, Jovanovich, 1958); Lennard Bickel, *The Deadly Element* (New York: Stein and Day, 1979).
7. T.C. Chamberlain, "On Lord Kelvin's Address on the Age of the Earth as an Abode Fitted for Life," *Annual Report of the Board of Regents of the Smithsonian Institution...for the Year Ending June 30, 1899* (Washington: Government Printing Office, 1901), p. 239. Reprinted from *Science*, n.s., vol. IX, no. 235, pp. 889-901, June 30, 1899, and vol. X, no. 236, pp. 11-18, July 7, 1899.
8. Rutherford's "playful suggestion" was recorded by a colleague, W.C.D. Whetham, and mentioned in a letter Whetham wrote to Rutherford on 26 July 1903, quoted in A.S. Eve, *Rutherford* (New York: Macmillan, 1939), p. 102.
9. *Brooklyn Daily Eagle*, "News Special," 30 August 1903, p. 1.

10. There is only one biography of Frederick Soddy, and it is somewhat idiosyncratic: Muriel Howorth, *Pioneer Research on the Atom* (London: New World's Publications, 1958). Additional sources are the works by Soddy cited below, and an issue on Soddy of the *British Journal for the History of Science*, vol. 12, no. 42, 1979. See especially Michael I. Freedman, "Frederick Soddy and the Practical Significance of Radioactive Matter," pp. 257-60, and Thaddeus J. Trenn, "The Central Role of Energy in Soddy's Holistic and Critical Approach to Nuclear Science, Economics, and Social Responsibility," pp. 261-276. While doing research on Soddy's views, one of the present authors (AJF) met Spencer Weart of the Center for the History of Physics, New York, who was engaged in similar research on Soddy and was also preparing a book which included discussions of Soddy's role in the early history of atomic energy. We exchanged draft manuscripts. Dr. Weart's study will be a major addition to these treatments of Soddy: Spencer Weart, *Nuclear Fear: A History of Images* (New York: Doubleday-Dial, forthcoming).

11. Lecture by Soddy to the Corps of Royal Engineers, Chatham, 1904; cited in Howorth, pp. 123, 283. Soddy published articles with similar comments during 1903 and 1904, also cited in Howorth, p. 238.

12. Frederick Soddy, *The Interpretation of Radium* (London: John Murray, 1909), p. 244. Again, similar comments were apparently in Soddy's lectures as early as 1903, according to Howorth.

13. Soddy, *Ibid.*, p. 229.

14. The first edition, published in March, 1909, was followed by enlarged editions in November 1909, 1912, and 1920. An expanded and updated version appeared as *Interpretation of the Atom* (London: John Murray, 1932), and again in greatly revised form as *The Story of Atomic Energy* (London: Nova Atlantis Publishing Co., 1949).

15. Hollis Godfrey, *The Man Who Ended War* (Boston: Little, Brown and Co., 1908).

16. Garrett P. Serviss, *A Columbus of Space* (New York: Appleton, 1911). Originally published as a serial in *All-Story Magazine*, January-June, 1909.

17. H.G. Wells, *The World Set Free: A Story of Mankind* (London: Macmillan, 1914).

18. *Ibid.*, p. 43. This is nearly a direct transcription from Soddy's *Interpretation of Radium*. The figure of "at least...a hundred and sixty tons of coal" was Soddy's revised estimate, appearing in later editions of *Interpretation of Radium*.

19. Arthur Train and Robert Williams Wood, *The Man Who Rocked the Earth* (Garden City, NY: Doubleday, Page, 1915. Rpt. New York: Arno Press, 1975).

20. Thirring, cited in note 4, above, p. 92.

21. Karel Čapek, *Krakatit*, originally published 1924, trans. Lawrence Hyde (New York: Macmillan, 1925).

22. Karel Čapek, *The Absolute at Large*, originally published 1922, trans. 1927 (New York: Macmillan, 1927).

23. Robert Nichols and Maurice Browne, "Wings over Europe," (New York: Covici – Friede, 1929). The original ending (apparently not performed)

had the young scientist killed in an accident before the assassination attempt.

24. John W. Campbell, Jr., "When the Atoms Failed," *Amazing Stories*, Jan. 1930.
25. Ernest Rutherford in *Nature*, 132 (16 September 1933), pp. 432-3.
26. This account of Leo Szilard's role is based on the collection of his recollections and correspondence edited by Spencer R. Weart and Gertrud Weiss Szilard, *Leo Szilard: His Version of the Facts* (Cambridge, Mass.: MIT Press, 1978). Szilard's comments on Wells begin on page 16.
27. *Ibid.*
28. *Ibid.*, p. 17.
29. *Ibid.*, p. 38.
30. For an excellent review of that critical work, see Hans G. Graetzer and David L. Anderson, *The Discovery of Nuclear Fission: A Documentary History* (New York: Van Nostrand Reinhold Co., 1971). The classic review paper was Louis A. Turner, "Nuclear Fission," *Reviews of Modern Physics* 12/1 (January 1940), pp. 1-29. Einstein's $E = mc^2$ is first used on page 7 of that review, and specifically identified on page 8. The equation is used to explain, in hindsight, why the work on fission was so important. A recent popular-level article reaches the same conclusions we have about Einstein's role: Robert Resnick, "Misconceptions about Einstein," *Journal of Chemical Education*, 57 (December 1980) 854-62.
31. Chapter 3 in Szilard's collection (note 26) discusses Szilard's version of these events. See also the account in Jungk, cited in note 6 above. Note that even this connection between Einstein and the bomb may have been given inflated importance. See Abraham Pais, *"Subtle is the Lord..."/the Science and the Life of Albert Einstein* (New York: Oxford University Press, 1982) p. 454 for a comment to this effect and additional citations.
32. As quoted by Jungk, note 6, p. 86.
33. George Bernard Shaw, ORT banquet speech, as included in note 35, below.
34. Rebecca West, "Blessed are the Pure in Heart," *Outlook and Independent*, 157/4 (28 January 1931), p. 156.
35. BBC-TV, *Einstein, The Story of the Man Told by His Friends*, 1969 release.
36. Author unidentified, "Crossroads," *Time* vol. XLVIII, no. 1 (1 July 1946), p. 52.
37. *Ibid.*
38. *Ibid.*, p. 56.
39. Joseph Laffan Morse, ed., *The Unicorn Book of 1952* (New York: Unicorn Books, Inc., 1953), p. 322.
40. Dennis Sanders, *The First of Everything* (New York: Delacorte Press, 1981), p. 56.
41. Robert Heinlein, "Blowups Happen," *Astounding Science Fiction*, September 1940. Rpt. in John W. Campbell, Jr. ed., *The Astounding Science Fiction Anthology* (New York: Simon and Schuster, 1952), pp. 1-42.
42. Theodore Sturgeon, "Artnan Process," *Astounding Science Fiction*, June 1941.
43. The actual separation of fissionable U235 from the much more common

isotope U238 was the major industrial challenge behind the development of atomic weapons. For a popular history, which illustrates the extensive and expensive corporate resources that were marshalled by the Manhattan Project for this purpose, see Stephane Groueff, *Manhattan Project* (Boston: Little, Brown and Co., 1967).

44. Anson MacDonald (pseud. for Robert Heinlein), "Solution Unsatisfactory," *Astounding Science Fiction*, May 1941.

45. Lester Del Rey, "Nerves," *Astounding Science Fiction*, September 1982. Rpt. in Raymond J. Healy and J. Francis McComas, *Adventures in Time and Space* (New York: Ballantine Books, 1975).

46. Cleve Cartmill, "Deadline," *Astounding Science Fiction*, 1944.

47. Clifford D. Simak, "Lobby," *Astounding Science Fiction*, April 1944.

48. Chan Davis, "The Nightmare," *Astounding Science Fiction*, May 1946. Rpt. in Groff Conklin, ed., *A Treasury of Science Fiction* (New York: Crown, 1948), pp. 3-18.

49. Pierre Boulle, "$E = mc^2$," in *Time Out of Mind*, trans. Xan Fielding and Elizabeth Abbott (New York: Vanguard Press, 1966. Original title: *Contes de L'Absurde suivis de $E = mc^2$*), pp. 288-352.

50. *Ibid.*, p. 290.

51. *Ibid.*, p. 352.

52. Klaus Volker, *Brecht: A Biography* (New York: The Seabury Press, 1978), p. 218.

53. Gerhard Szczesny, *The Case Against Bertolt Brecht*, trans. Alexander Gode (New York: Frederick Ungar, 1969), p. 14. For additional discussion, see Eric Bentley's introduction in *Bertolt Brecht*, "Galileo," English version by Charles Laughton, ed. Eric Bentley (New York: Grove Press, Inc., 1966).

54. Friedrich Dürrenmatt, *The Physicists*, trans. James Kirkup (New York: Grove Press, Inc., 1964; orig. *Die Physiker*, 1962).

55. Cited in note 6.

56. Dürrenmatt, p. 84.

57. Dürrenmatt, p. 94.

58. "Target One," a radio play on the series *X Minus One*, originally broadcast 26 December 1957 (?). Quotations from a rebroadcast on KSFO radio, San Francisco, CA, 10 April 1974. No other documentation on this particular program has been found.

59. See note 90, Chapter 4.

60. Wells, note 17, p. 233.

61. Dürrenmatt, note 54, p. 89.

62. Lansing Lamont, *Day of Trinity*, (New York: Atheneum, 1965), p. 257. This book offers an addition to the Einstein myth by recounting a wartime anecdote about Einstein being at Los Alamos [an event which did not occur].

63. "Purdue Opinion Panel" in H.H. Remmers and D.H. Redler, *American Teenager* (Indianapolis: Bobbs-Merrill, 1957). See also *Attitudes Towards Science*, Report 08-5-02 (Denver: National Assessment of Educational Progress, 1979).

64. The stamp shown in this figure was collected by Dr. Arthur Luehrman of Berkeley, CA, in Beijing, People's Republic of China, in 1979.
65. Cover of *Look*, 1/4, 2 April 1979.
66. Italian advertisement for Carlsberg beer, 1970's. Collected by Judith Goodstein, California Institute of Technology.
67. A plausible account of the origin of this legend is presented by Stephen G. Brush, "Einstein and Indeterminism," *Journal of the Washington Academy of Sciences*, 69/3 (1979), p. 90.
68. George Mampilli, "Einstein and Maya," *The Illustrated Weekly of India*, vol. C10, 11-17 March 1978.
69. Albert Einstein, *Uber die Spezielle und die Allgemeine Relativitatstheorie, Gemeinverstandlich* (Braunschweig: Vieweg, 1917). The authorized English translation of this popularization appeared as *Relativity, the Special and General Theory: A Popular Exposition*, trans. Robert W. Lawson (London: Methuen, 1920). The book has remained in print continuously, and Einstein added to it from time to time. The fifteenth edition was completed in 1952.
70. Advertisement in the *Daily Californian*, 2 June 1975, p. 9.
71. Donald T. Lunde, *Murder and Madness* (San Francisco Book Co., 1976), p. 65.
72. *Ibid.*, p. 76.
73. Nicholas Tomalin and Ron Hall, *The Strange Last Voyage of Donald Crowhurst* (Briarcliff Manor, N.Y.: Stein and Day, 1970). Paperback edition, 1979. Page numbers refer to this edition.
74. *Ibid.*, p. 237.
75. *Ibid.*, p. 238.

INDEX